Toward Effective Strategic Analysis

Also of Interest

Arms Control and Defense Postures in the 1980s, edited by Richard Burt

Expert-Generated Data: Applications in International Affairs, edited by Gerald W. Hopple and James A. Kuhlman

Verification and SALT: The Challenge of Strategic Deception, edited by William C. Potter

Quantitative Approaches to Political Intelligence: The CIA Experience, edited by Richards J. Heuer, Jr.

*China, the Soviet Union, and the West: Strategic and Political Dimensions for the 1980s, edited by Douglas T. Stuart and William T. Tow

*NATO—The Next Thirty Years, edited by Kenneth A. Myers

The Evolution of U.S. Army Nuclear Doctrine, 1945–1980, John P. Rose

*Available in hardcover and paperback.

Westview Special Studies in National Security and Defense Policy

Toward Effective Strategic Analysis: New Applications of Information Technology

Albert Clarkson

Exploring the future of strategic analysis, this book identifies problems at the heart of the historical U.S. failure to perform effective strategic analysis, then explains and dramatizes how new applications of information technology can make significant progress possible. Certain specific limitations of human memory, says the author, are major causes of our inability to develop and maintain the analytic contexts of models, to see meaning in incoming data, to conduct and learn from appropriate *post mortems* that measure analytic effectiveness, and most important, to develop a powerful strategic imagination. In a call to action to analysts, technologists, scientists, and managers, he shows that effective analysis requires the development of specialized extrasomatic machine memories—memories that will aid the indispensable analyst in improving the art of analysis while facing a growing strategic imperative.

Firmly grounded in considerable field experience in strategic analysis, the book draws ideas also from contemporary research in cognitive psychology, projection techniques, information science, literature, and other fields.

Albert Clarkson is manager of decision research and technology at ESL, Inc., where he has directed a number of projects concerned with national defense and security problems. He formerly was head of the strategic research department at ESL.

for
ELEANOR

Toward Effective Strategic Analysis: New Applications of Information Technology

Albert Clarkson

Westview Press / Boulder, Colorado

Portions of the research underlying this book are based on work done under contract to the Defense Advanced Research Projects Agency. The views and conclusions contained in the book are those of the author and should not be interpreted as necessarily representing the official policies, either expressed or implied, of the Defense Advanced Research Projects Agency or the U.S. Government.

Westview Special Studies in National Security and Defense Policy

All rights reserved. No part of this publication may be reproduced or transmitted in any form or by any means, electronic or mechanical, including photocopy, recording, or any information storage and retrieval system, without permission in writing from the publisher.

Copyright © 1981 by Westview Press, Inc.

Published in 1981 in the United States of America by
  Westview Press, Inc.
  5500 Central Avenue
  Boulder, Colorado 80301
  Frederick A. Praeger, Publisher

Library of Congress Catalog Card Number: 81-69202
ISBN: 0-86531-243-5

Composition for this book was provided by the author.
Printed and bound in the United States of America.

Contents

List of Tables ix
List of Figures ix
Author's Preface xi

1 INTRODUCTION 1

2 STRATEGIC ANALYSIS MODEL 7

   Profile of the Strategic Analyst 7
   A Definition of Strategic Analysis. 9
   The Settings of Strategic Analysis. 10
   Stages of Strategic Analysis. 11
   Themes Sounded in the Discussion of the Model 19

3 STRATEGIC ANALYSIS MEASURES 21

   The Conceptual Framework 22
      The Essential Dynamic of Analysis. 22
      Perspectives from Living Systems Theory. 24
      Perspectives from Information Theory. 24
   Basic Measures 26
   Effectiveness 29
      Obstacles to Effectiveness. 30
      Epistemological Obstacles. 32
      Preliminary Framework for Effectiveness Measures 35
   Readiness. 35
      Obstacles to Readiness 36
      Preliminary Concepts of Readiness Measures. 37
   Maintenance. 37
      Obstacles to Maintenance 38
      Preliminary Concepts of Maintenance Measures. 39

Key Information-Related Variables. 39
Backlog. 39
Relationality 40
Signification. 41
   The Problem of Real Effectiveness 41
   Levels of Signification 44
A Symbolic Representation of Strategic Analysis. 47
Transition 54
Some Basic Problems of Measurement. 56
   Problems of Real and Estimated Effectiveness. 57
   A Viewpoint on Estimated Effectiveness 68
The Measures and Variables as a System. 71
   Commentary on Table 4. 71
   Subvariables. 76
   States and Pathologies 77
Some Comments on the Uses of Measurement Data 78
Transition 79

4   THE ART OF STRATEGIC ANALYSIS 81

Major Problems. 81
Some Features of the Proposed Cognitive Technology 83
Monitoring Stage. 86
Threat Recognition Stage 87
   Analysis Forms. 88
   Analysis Routine. 94
   Signification. 100
   Cognitive Technology Implications. 103
Projection Stage 105
   Analysis Forms. 105
   Analysis Routine. 126
   Signification. 126
   Cognitive Technology Implications. 134

5   THE STRATEGIC ANALYST IN THE FUTURE:
      SPECULATIONS 141

SOURCE NOTES 163

SELECTED BIBLIOGRAPHY. 173

INDEX 177

Tables and Figures

Tables

1	Functional Outline of Monitoring Stage	13
2	Functional Outline of Threat Recognition Stage	15
3	Functional Outline of Projection Stage	17
4	Interrelationships Among Measures and Variables	72
5	Measures Linkage	76
6	Example Use of Information Categories with Threat Models	109
7	Example Specificity Rating Scale	110
8	Examples of Sectors	114
9	Example Rationale Matrix	125
10	Example Rationale Formats	127

Figures

1	Basic Concept of Computer-Based Strategic Analysis	8
2	Major Functional Stages in Strategic Analysis	12
3	Measures Emphasis	23
4	Strategic Analysis and Elements of Living Systems Structure	25
5	Basic Measures Diagram	27
6	Hierarchy of Measures	28
7	A Symbolic Representation of Strategic Analysis	50
8	Methods Emphasis	85
9	Example of PAMNACS	90
10	Second Example of PAMNACS	91
11	Example of DEN	92
12	Second Example of DEN	93
13	Example CEF	95

14	Single-Sector Map – Military	116
15	Single-Sector Map – Economic	117
16	Single-Sector Map – Internal Security	118
17	Single-Sector Map – Leadership Outlook	119
18	Example Probable Futures Map	121
19	Design Elements	136

Preface

No one writes a book alone. The cooperation, support and wisdom of others are necessary gifts for a writer. To several people I will never be able adequately to express thanks. As you learn when you write, words often fail. But however much words fall short, I must offer words of thanks to several people for their crucial help.

The research which led to this book would not have been possible without the support of Judith Ayres Daly of the Defense Advanced Research Projects Agency. A pioneer in the development of computer-based analysis support systems, Ms. Daly has been a major designer and sponsor of cognitive technology over the past several years. It has not been an easy endeavor; there have been long, gray days as well as some bright ones. I owe her many thanks.

I could not have written the book without the research opportunities made possible by Captain Thomas Curry, USN, to whom I am grateful.

I thank my friends, mentors and colleagues, John W. Sutherland, Jerry S. Kidd, Thomas G. Belden, and R. Jack Smith for pursuing with me over several years many critical discussions on several aspects of the research. Their help has been indispensable.

I owe a considerable debt to many others as well, friends and colleagues who offered insights, sound advice and encouragement; and who bear no responsibility for failures and weaknesses in the research. Among these friends are: Jack Omen; Frances Calaway; Captain John Dillon, USN (Retired); Gerald L. Fuller; Don R. Harris; Richards J. Heuer, Jr.; Vince Heyman; Gerald W. Hopple; Laurance Krasno; Joshua Lederberg; Colonel R. E. Littlefield, USA; Benny Meyer; James Grier Miller; George Nisihara; William J. Phillips; Captain Frederick W. Robitschek, Jr., USAF; Captain Clyde Smith, USN; and Malcolm Stewart.

Thanks to several people I have been fortunate to know during earlier periods of my life, people of skill and character who taught me much: Jean A. Davis; Harvey G. Davis, Sr.; William Brown; William J. Perry; Jan Kozeluh; and Robert O. Bowen.

Thanks to Lynne Rienner of Westview Press for several excellent suggestions for textual refinements and other modifications.

I wish to thank Lew Franklin and Jean Isabeau of ESL Incorporated, for their support. For several years I have been a member of the technical staff of ESL, a pioneering company that continues to be remarkably successful in high technology applications. More than I can say, the experience and support I have gained there have been and continue to be invaluable.

Special thanks to Jo Franklin, Terri Monk and Cheryl L. Wright who patiently drafted the many versions of the manuscript; and to Dick Lee for his illustrated vision of the strategic analyst at work.

A.C.
Los Gatos, California

1
Introduction

The present book is about the future of strategic analysis. Strategic analysis has many forms: it seeks to anticipate foreign situations that will critically influence a nation; to envision conditions which will decisively affect an economy or an industry; to grasp the likely consequences of current policies of a government; and to accomplish many other objectives which serve the growing imperative in our time to see ahead, to reconnoiter the future. We are learning all too painfully that unless we become more successful at strategic analysis, our plans, policies and decisions will become more absurd and pathetic; surprises more dangerous; the sense of futility more intense; and the issue of accountability more difficult.

Yet obviously strategic analysis continues to be practiced often with little, if any, success in government departments and agencies; private institutions; corporations and businesses; and the academic, artistic and journalistic communities. Certainly at this moment, at the beginning of the Eighties, none of us can do it well. Luckily perhaps, but not well, and therefore not convincingly. We are all amateurs.

Yet within the past few years the future of strategic analysis has seemed to me to become very bright. For with the advent of cognitive technology, man has at last begun to develop a weapon with which he may begin a long, arduous but promising struggle to mitigate the modern crisis of interpretation. This promise of greater success in strategic analysis does not lie precisely in better data processing, smaller and cheaper computers, computer-based expert systems or artificial intelligence, although these developments are exceedingly important and rightfully lauded today. But it *does* lie in our discovery and invention of increasingly refined information machines, and especially in our growing sense of their tremendous potential.

In certain crucial senses, we humans have not yet taken to information machines. For reasons that will be discussed in later chapters,

2 TOWARD EFFECTIVE STRATEGIC ANALYSIS

I seriously question that we have yet begun really to work with them. I sense, however, that we are ready. And I am convinced that when we do we will transcend certain dramatic cognitive limitations which now, as they have for thousands of years, virtually preclude effective strategic analysis. Just as automobiles and airplanes suddenly appeared and mitigated spatial and temporal limits, so information machines must mitigate human cognitive limits; not merely limits on calculation and storage, but limits on imagining, on the development of realistic models — schemata or knowledge structures — and on successful inferencing. And indeed, the "man-machine relationship" here will be a close, highly personal one.

Moreover, it will be a relationship between the machine and the artist. Man proceeds through the world not primarily as a scientist but primarily as an artist. In strategic analysis there is no magic algorithm. Because we are interpreting a reality which by definition will bring novel situations, there are few safe assumptions of replication in any sense. Hence though it must be as rigorous as possible, and understood, measured and developed as scientifically as possible; and though it must include techniques based in quantification and in probability, strategic analysis is finally an art. Arguably it is becoming our most important one.

The strategic analyst will realize a more exalted art, pursuing literal realism about the future more successfully if still, of course, imperfectly, only if we make the machine truly extend his cognitive powers. The machine must become "extrasomatic": it must become almost a part of him. But if the machine becomes crucial, it will also remain subordinate: it will facilitate, rather than replace, the human analytic functions, for these are irreplaceable.

Most importantly, the machine must extend the analyst's memory. The strategic analyst must have an extrasomatic memory.

Several factors, elaborated in subsequent chapters, have led me to this perspective.

To begin with, in the view of many there is now a growing epistemological crisis. Perhaps unprecedented in its demoralizing effect, this crisis has thrown into doubt our interpretive ability, our success in making sense of events, our ability to project outcomes. There is a deep pessimism in many well-informed, well-intentioned people about prospects for improving human capabilities to look ahead, anticipate crises, and develop enlightened policies. Theories, both formal and informal, assert that among the root causes of our difficulties are certain very old psychophysiological constraints on the human mind. Moreover, such root causes are perceived as being continually and increasingly exacerbated by the modern information deluge (perceived now as an "overload"), a sort of huge and peculiar collagé added to daily by the contemporary communications media but perversely discontinuous. While it promises much, the multitudinous data actually frustrates our interpretive impulse and conditions us to limited perspectives. In fact, it tends to cut off our inquiry into these problems because it tends to anesthetize us.

In this context, several constraints on human cognition have now become crucial to the problem of strategic analysis. The most fundamental is man's *difficulty in reconstructing his past analytic perspectives.* In greatly simplified terms, it is a "failure" of memory, a quiet but all-important discontinuity, which we experience as we perform cognition through time; we forget as we learn. Richards J. Heuer, Jr., has written with considerable insight about this problem as it affects intelligence analysis.

In turn, several related problems arise from the constraints on human memory. The first consists of *limitations on both the scope and intensity of modeling.* In strategic analysis it is vital to create complex frames of reference. These arise from man's essential strategy of building models of past, present and future reality. This framework of models, an analytic context essential to the discovery of meaning and to learning, will always be fundamental to strategic analysis. In our era of intense communications media and a deluge of information, the problems, imperatives and promises of creating interpretive contexts seem larger than ever. Yet we remain profoundly deficient in creating *operational* contexts. Given that analytic energies are not only precious but difficult to sustain, both the building of, and the analysis against, complex contexts is severely constrained in the unaided mind.

An ensuing constraint is *the difficulty of the mind in rapidly perceiving ramifications in meaning through complex frameworks of coherent perspectives — through, in short, many assumptions and elements of logic — that occur when individual judgments are made about small portions of the overall framework.* This problem of our limited ability to grasp implications assumes, of course, that we have been able to construct sophisticated models to begin with! Obviously we need effective contexts both for developing meaning in strategic analysis and to mitigate the information overload.

A further constraint, also arising from the problems of human memory, is *our difficulty in measuring the effectiveness of strategic analysis.* In essence, human problems in reconstructing past analytic perspectives impair *post mortems*, thereby curtailing diagnosis, prescription and, finally, learning. I shall discuss in detail the difficulties and paradoxes in attempting to measure the effectiveness of strategic analysis, but I must note now that our prospects for developing a tradition of increasingly successful strategic analysts are strongly tied to the need for standards, for measures: these are basic to the development of both motivation and mastery.

These and other fundamental obstacles have led me to solutions that involve computer technology. If we remember that the human analyst is limited; that the machine, if differently, is much more limited; and that the strategic analyst finds himself without choice engaged in an heuristic process, then it seems evident that the most important immediate question becomes: how should we aid the now largely unaided human mind in performing strategic analysis? Stated very simply, I argue that we must begin to use our ingenuity to create computer-based extrasomatic memories of analytic operations, memories which in their various forms also hold the analytic context developed heuristically by the analysts, themselves. The machines will allow us to record and thereby reconstruct past analytic perspectives, safely removed from the losses of human memory. At the same time, the machines will allow us — indeed, must prompt us — to build and preserve as well as review, criticize, dismantle and recreate, the vital context for making strategic analytic judgments. And indeed, for many reasons the analytic context I speak of as being stored extrasomatically must complement the schemata (or knowledge structures) of the mind, those mental constructs which allow us to recognize portions of reality, interpolate and extrapolate, infer and perform other mental operations. And in the final analysis, the key factor becomes the *strategic imagination*.

That is indeed what much of this book ultimately is about: the strategic imagination, its development and future. We shall explore these problems at length.

I should point out that there are specific research and development implications in many of the discussions in later chapters. To remove the need to refer to them frequently, I will indicate them here. Although this book is written primarily from the viewpoint of a technologist, the insights and research of cognitive psychologists and other scientists such as Amos Tversky, Richard Nisbett and Lee Ross seem to me fundamental to the issue at hand. In designing cognitive technology, we must utilize the findings of cognitive psychologists: their recent experiments and research into judgmental heuristics, knowledge structures and other aspects of human cognition have already provided valuable insights. Indeed, as I indicate in later discussions, it is a hope of mine that this research and experimentation can be brought closer to the world of the strategic analyst, and that the new memories I discuss will be one of the catalysts for this development. I must apologize now for any failures to represent adequately the findings and insights of cognitive scientists — my colleague, Richards Heuer, is currently exploring these findings and defining applications of them to analytic art with precision — and hope I will be forgiven any errors in seeking to summarize the results of experimentation.

I also believe that some of the methods and procedures of forecasting — for example, certain analytic approaches developed by Willis Harman — are of considerable potential value to strategic analysis. If there has been a tendency by some to dismiss such methodologies, I would suggest that the pursuit of the methodologies with the aid of cognitive technology may greatly enhance the value of the methods.

Finally, the quest for, and attainment of, order and system that mark the work of the successful systems scientist are important to the development of a tradition of effective strategic analysis. I have been fortunate to be able to discuss some of the complexities of strategic analysis with James Grier Miller and to count among my co-workers on strategic analysis projects John W. Sutherland; these two men are clearly among the finest of systems scientists, and their skills and insights have brought home to me many times the key future role of their field in the development of strategic analysis.

In the chapters which follow, I shall discuss these and other aspects in detail. In Chapter 2, the process of strategic analysis is considered. I have developed a functional model of analysis, dividing the process into interrelated stages and identifying procedures and steps associated with each. In Chapter 3, I pursue the issue of effectiveness in strategic analysis, developing a system of interconnected measures of analytic effectiveness which I believe to be a new and powerful approach to a difficult problem which has confronted us for some time. In Chapter 4, I explore the art of strategic analysis and offer designs of cognitive technology to support it. An entire system of computer-based strategic art is described. Finally, in Chapter 5, I offer some speculations about strategic analysts of the future, discussing innovation, aesthetics and other aspects.

I should note that this is a book in which the explorations and definitions of ideas such as "strategic analysis," "strategic analyst," "cognitive technology" and "analytic effectiveness" are gradual and cumulative. There are, for example, several portraits of the strategic analyst, beginning in Chapter 2 and culminating in Chapter 5. Similarly, the idea of strategic analysis is elaborated slowly through the text because I have found no adequate way to convey quickly the complexities of the concept.

Finally, the reader will see that my experience lies chiefly in work on problems of defense and national security. But I am primarily interested in the "generic" strategic analyst of the future. Such analysts will work in government, the corporate sector and a variety of other settings. While I do not foresee a revolutionary impact for the new strategic art and its supporting technology in the next several years, I do see now the beginnings of a significant movement in that direction. With this book I hope to give some impetus to that movement.

Note: Source notes and references are to be found at the rear of the book and are keyed to page numbers and phrases/sentence fragments.

2
Strategic Analysis Model

The first task is to develop a useful model of strategic analysis. Given the encouraging contemporary progress in the understanding and modeling of human cognition, we nevertheless remain far too ignorant to make any pretense of modeling all activities in a mental process as elusive and complex as strategic analysis. We must be content with a functional model which identifies major stages, procedures and decision points. From there we must identify crucial variables in the cognitive process which can be monitored, change measurably, and are diagnostic. Such a model is essential in developing analysis performance measures.

Better, more sophisticated models of the analysis process constitute a goal. The present model is only a start.

Figure 1 shows an overall concept of computer-based strategic analysis and the particular area of emphasis of the present discussion. The elements in the figure are explained in this and subsequent chapters.

Profile of the Strategic Analyst

We should begin with a brief description of the strategic analyst. We will enlarge the portrait in later chapters.

The strategic analyst watches some part of the world — an economy, an industry, a foreign country. He receives information of many kinds and levels of credibility from many sources on events and situations in his sphere. He researches problems and prospects of leaders, countries, economic sectors, alliances, and regions. Perhaps he has visited his assigned part of the world. If responsible for a foreign activity, he may have learned one or more of the languages of the area. He has undergone selected training.

The strategic analyst is supposed to do no less than this: understand the dynamics and prospects of change in the area he watches; recognize

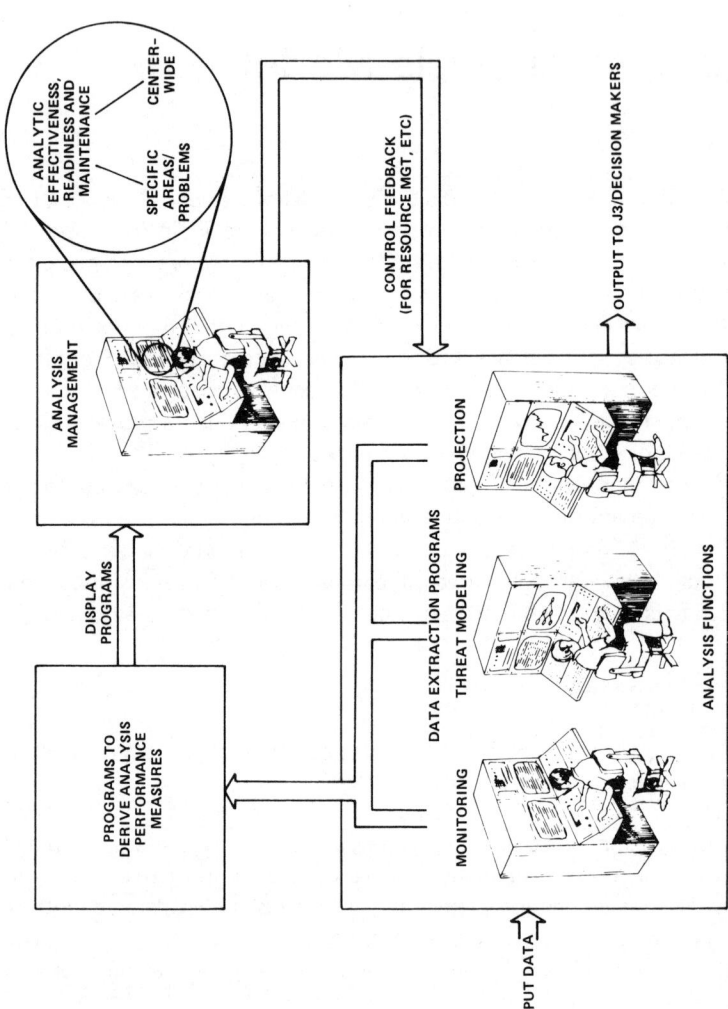

Figure 1. Basic Concept of Computer-Based Strategic Analysis.

signs of change, particularly threatening change; assess its significance; make projections; and carry out warning as appropriate.

He strives for literal realism but properly acts on the basis of probabilities.

Through modern communications technology, he encounters on the job (and off the job, too) an unprecedented, growing, vast amount of information, much of it individual facts, isolated events. He is one of the people most caught up in the information revolution. The myriad data provides diverse, sometimes conflicting views of parts of reality.

He is within the realistic tradition of philosophy. He is an empiricist. His operational epistemological assumption is that in some sense "reality" is objective. It is knowable. There are at least some correspondences between his experience and the facts of reality. Moreover, the world in its change has an order, governed by sets of laws, some of which are now known and can be understood and recognized by humans in their perception of reality.

His chief impulse is interpretive. The strategic analyst will be seeking to order the chaos of experience, much of that experience being secondhand, its objects removed like echos from the scene of his interest. He will seek to discover as much certainty and necessity within the reality he watches as possible. To some extent, he must believe that changes are comprehensible in terms of explainable cause and effect. If change is perverse, it is also consistent within some bounds; it is similar to some degree across different times and places. This basic logic of change derives from the assumptions of purpose and of reality governed by laws recognizable by the rational mind operating on current and past information. Hence factual data are taken as reliable cognitive elements.

Perhaps most fundamentally, he will believe implicitly in the authenticity of models organized by "plots"; models in which experience is ordered in chronological, causal forms. This is not to exclude other perceptual and analytic modes such as mosaic techniques.

Analyzing input data with his models and analytic techniques, perhaps he can be said to have two modes of interpretation:

- To extract a pattern of significance from information
- To impose a scheme of meaning on reality.

A Definition of Strategic Analysis

Implicit in the profile of the strategic analyst is a definition of strategic analysis. Stated explicitly, it is as follows: Strategic analysis is a rigorous cognitive process by which possible crucial realities of the future are first imagined and then modeled systematically to delineate their conditions, dynamics and potential outcomes, every effort being made to achieve realism and verisimilitude; with various inferential strategies procedurally employed to develop comparative probabilities; with the models and probabilities continuously reviewed and modified appropriately on the basis of new data; with *post mortems* conducted systematically to measure performance and to stimulate learning; and the entire process oriented toward decision making and policy formulation.

I will elaborate considerably on the various elements in this definition as we pursue several discussions below. For now, I shall comment briefly on the definition by noting that strategic analysis:

- Involves projections
- Is concerned with capabilities and intentions of societal systems such as nations and other organized entities
- Is concerned with political, economic, military and cultural dimensions
- Is oriented toward time-critical monitoring of current events
- Also involves a long-range estimative orientation
- Is a great challenge, there being enormous epistemological, cognitive, institutional and other constraints on its effectiveness.

But I am most concerned with strategic analysis as a *cognitive process*. The model of this process described below reflects part of my understanding of strategic analysis as observed during protracted research at various analysis centers. The model also includes stages and procedures not necessarily pursued formally now in the manner described, but which analysts and managers have looked upon as necessary and promising.

The Settings of Strategic Analysis

The strategic analyst performs long-range planning, forecasting and warning, projections and estimates, and other tasks. He does so in a variety of settings. He may do analysis in government analysis centers in which there are various analysts responsible for such functions as watch; indicator analysis; long-range estimative analysis; and contingency planning. In some fashion, usually similar to the above task breakdown, such analysis organizations are typified by bureaucratic arrangements in which groups are established to perform the various functions of analysis, beginning with review of incoming data and concluding with policy definition and review.

Interestingly, the number of strategic analysis staffs in the corporate world appears to be growing. Corporations are also beginning to rely more heavily on external consultants and service organizations for assessment of political and economic risks associated with various decision options. We must anticipate that the corporate world will become a major setting of strategic analysis. Certainly the discussion of strategic analysis in the present book is meant to apply not only to intelligence and national security analysts, planners and managers; but to their counterparts in private industry as well.

Stages of Strategic Analysis

As a first step toward modeling strategic analysis as a process, Figure 2 shows what I consider to be the major progressive, interrelated functional stages in strategic analysis: *monitoring*, *threat recognition*, and *projection*.

The *monitoring stage*, outlined in Table 1, may be viewed as a sequence of steps and procedures beginning with the review of incoming data, followed by the sorting of the data into different categories of interest; a comparison of the data with predetermined indicators whose activity is judged to have significance in terms of affecting plans, impacting on standing decisions, and signifying emerging threats; an identification and display of correlations; a preliminary evaluation and specification of the significance of the activity; and the movement of the results to the next stage of analysis.

Note that steps 8 and 9 comprise a safeguard procedure against analyst tendencies to overlook novel situations. The strategic analyst must guard against cognitive biases and problems with information which may impede his recognizing the onset of such conditions. Strategic analysts should be alert to "ambiguous" and "anomalous" data since it may actually signify a crucial development or situation not previously well-imagined and modeled by an analytic community. The problem of novel situations is, of course, a fundamental challenge. Steps 8 and 9 begin a safeguard routine within the analytic procedures which extends into later stages.

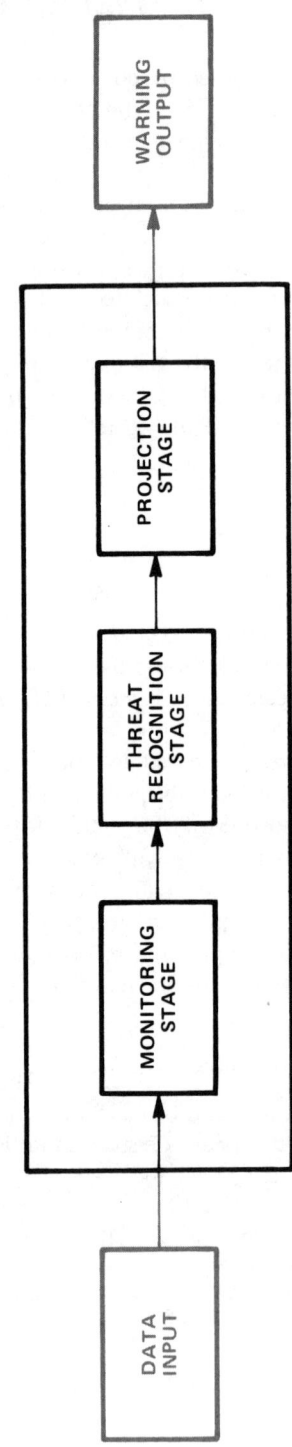

Figure 2. Major Functional Stages in Strategic Analysis.

Table 1. Functional Outline of Monitoring Stage

Step 1: Review incoming data

Step 2: Sort incoming data

    Procedure 1: Geographic
    Procedure 2: Military
    Procedure 3: Political
    Procedure 4: Economic
    Procedure 5: Cultural
    Procedure 6: Other

Step 3: Match sorted data against predetermined indicators

Step 4: Identity correlations

Step 5: Display correlations

Step 6: Make preliminary evaluation of indicators

    Procedure 1: Assess relative levels of activity
    Proecdure 2: Assign significance to activities

Step 7: Route results to next analysis stage

Step 8: Review residual data

Step 9: Move appropriate residual data forward for further analytic review

14 TOWARD EFFECTIVE STRATEGIC ANALYSIS

The *threat recognition* stage begins where the monitoring stage leaves off. The steps and procedures are described in Table 2.

The analyst now progresses from matching incoming data with individual indicators to associating active indicators with models of situations ("scenarios") previously developed in some chronological and causal form. This is a higher level of cognitive challenge: some indicators — at early enough stages perhaps a sizeable portion of them — may seem capable of being associated with roughly equal probabilities to more than one possible threat.

The future cannot, of course, be precisely modeled. The fundamental risk that novel situations will occur whose shape and dimensions are not recognizably analogous to models currently in an analyst's inventory — and, indeed, may be obscured by undue focus on existing models — is omnipresent. Yet models abound. They have been developed in many forms and levels of detail many times in the history of strategic analysis and will continue to appear. Surely all analytic organizations of any magnitude and sophistication have constructed scenarios depicting how adverse actions might arise or be instigated by opponents. Models are both imperfect and a valid and necessary response to the challenges facing the strategic analyst. The important ideas are that models be viewed as approximations and treated with appropriate skepticism, and that in an era of information technology they be constructed in useful, flexible forms (e.g., rapidly displayable and changeable). (In subsequent chapters, considerable attention will be focused on these issues.)

The *projection stage* represents the final and by far the most formidable analytic challenge. The analyst now surpasses his role as historian and explores the future. The making of projections is a pervasive and highly refractory enterprise. From weather forecasting through human resources planning to economic prognostication, it is rarely done to the satisfaction of the forecaster or the consumer of such forecasts. Nonetheless, I confidently advance a set of procedures for threat projection that seem to me to be thorough, easily usable, and useful to analysts. This set of procedures is described in detail in Chapter 4.

We may distinguish four types of projections significant in strategic analysis. First, projections may take two forms: *predictions* and *forecasts*. By my definition, prediction involves an estimate by analysts of conditions necessary for given situations. However, predictions are not necessarily held to be likely (or very likely) by analysts. They will, however, tend to

Table 2. Functional Outline of Threat Recognition Stage

Step 1: Match indicators from monitoring stage against predetermined threat models

   Procedure 1: Call up preestablished threat models
   Procedure 2: Match indicators against models

      a. Call out critical events/stages (or nodes) of threat models

      b. Search for correlations to indicators

      c. Identify correlations

      d. Review correlations against other critical events whose occurrence has been previously detected within developing situational context

      e. Assign confidence values to results of c and d

      f. Identify key events and activities whose occurrences have not been detected

      g. Identify further information needs

      h. Compare all preestablished threat models to which the data has correlated; establish:

         (1) Duplications
         (2) Similarities
         (3) Differences

      i. Explore for possible alternative explanations/hypotheses for data

Step 2: Move pertinent threat models to projection stage of analysis

Step 3: Conduct novel threat analysis

   Procedure 1: Formulate new threat hypotheses
   Procedure 2: Display threat hypotheses
   Procedure 3: Compare residual data against hypotheses

Step 4: Move appropriate novel threat models to projection stage of analysis

address situations of great interest from the strategic viewpoint. A practical view of predictions is to consider them a vehicle of analytic preparation. In the process of forming and refining predictions, the analyst thinks about (indeed, *analyzes*) various existing, imminent and/or remote conditions whose occurrence could create threats.

The term, forecast, is used to refer to a projection in which an analyst estimates that the event or situation *will* occur. Hence a forecast may be thought of as a prediction which the analyst now believes will in fact take place with an implicit or explicit probability attached.

Further, both predictions and forecasts may be of two types: *unimpeded* and *influenced*. An unimpeded projection, whether a prediction or a forecast, is one in which the analyst judges that under the current policies and posture of the friendly side, the projected situation is likely to occur in a certain form. In other words, the analyst judges that unless U.S./friendly decision makers intervene and change conditions over which they are presumed to have control, the projected outcome is likely to take a certain form, designated the unimpeded projection. The influenced projection is one in which the analyst considers the impact of certain activities which his decision maker may introduce and their impact on the outcome. This becomes the influenced projection.

Table 3 is an outline of functional procedures and steps in the projection stage. Most of the procedures are self-explanatory (and will be elaborated in later discussions of methodology), but comments are appropriate for procedures 1-3 under step 1. For each of the four types of projection, the first procedure entails an important transformation: the analyst translates the individual threat models from the preceding stage into a predictive format organized according to the basic information categories — who, what, where, when, how, and why. In my judgment, projections should ultimately be expressed in a very fundamental grammar, namely, the information categories linked together in English sentence structure. This grammar of projections can accommodate simple or complex projections. (Examples are given in Chapter 4 of the transformation of situation models into this format.)

The second step entails comparing the specificities of individual threat projections. A Specificity Rating Scale by which analysts may estimate the degree of specificity for each of the information categories in any given threat projection has been developed and is described in Chapter 4. Predictions and forecasts expressed formally in terms of who, what, when, where, how, and why may be compared quantitatively in each category. For example, the category, *when*, may be highly specific

Table 3. Functional Outline of Projection Stage

Step 1: Compare candidate threats in unimpeded, predictive domain

   Procedure 1: Translate individual threats into predictive format based on information categories

     a. *Who* (or what)
     b. (Could do) *what*
     c. (to) *whom* (or what)
     d. *Where*
     e. *How*
     f. *When*
     g. *Why* (or because X conditions apply)

   Procedure 2: Compare specificities of individual threat predictions

   Procedure 3: Elucidate rationale behind specificities

     a. Estimate conditions that would support various threat enactments, and their relative importance and interdependence

        (1) Political situation

        (2) Military capabilities

        (3) Economic conditions

        (4) Social/cultural conditions and forces

        (5) Decision maker/leadership perceptions and predispositions

        (6) Other

     b. If conditions now hold, estimate earliest possible occurrence in future of threat enactment

     c. Identify any missing conditions

     d. For missing conditions, estimate earliest occurrence in future

     e. Estimate earliest point in future pertinent threats could occur after missing conditions occur

Table 3 (cont'd)

Procedure 4: Indicate sources (e.g., outside expertise)

Procedure 5: Indicate/summarize assumptions

Procedure 6: Indicate/summarize uncertainties and data requirements

Procedure 7: Across predictions, and added to previous comparisons, isolate and identify by information category the following:
  a. Duplications
  b. Similarities
  c. Differences

Procedure 8: Compare predictions in terms of probability against selected timelines

Step 2: Compare candidate threats in influenced predictive domain

Procedure 1: Consider potential U.S. and friendly decision maker influences on development of situations
  a. Identify friendly influence options

Procedure 2: Translate influenced threat models into predictive format (as in step 1 above)

Procedure 3: Repeat rest of procedures in step 1

Procedure 4: Make grand comparison of unimpeded and influenced predictions

Step 3: Perform similar procedures for unimpeded and influenced forecasts

(described, say, to the hour) or extremely general.

The third procedure for each type of projection is to develop and elucidate the rationale. This, of course, is the greatest challenge to the analyst, the most difficult single analytic challenge in the entire process of performing strategic analysis. The analyst must analyze the conditions that might lead to various threat enactments and assess the relative importance and interdependence of the conditions. The conditions could include military capabilities, political aspects, economic factors, social and cultural conditions and forces, decision maker/leadership perceptions and predispositions, and other key variables. This third procedure must receive the greatest emphasis in methodology.

Themes Sounded in the Discussion of the Model

Several very basic themes in the strategic analysis problem have been sounded which will be discussed further in subsequent chapters. These include: limitations in understanding and modeling human cognition; cognitive and other obstacles to analytic effectiveness; the relationship of information technology to advanced strategic analysis; the methodological crisis; and the problems of monitoring and managing analysis. A separate theme is the expository difficulty in conveying the actual dynamics, excitement and incentive in performing complete strategic analysis, as modeled above. But a major theme is this: a serious pathology of strategic analysis is the failure to *complete* the full process of analysis.

All these issues — and others as well — are at the heart of the next problem we must confront: the development of analysis performance measures.

3
Strategic Analysis Measures

Two points should be made at the outset. First, we will explore and develop performance measures for strategic analysis without direct concern about how they would be implemented. The primary purpose at this point is a conceptual analysis of the measurement problem. Implementation is described in detail in Chapter 4.

Second, the overall system of measures should be previewed. It is a comparatively simple system. There are three major (and classic) measures: analytic *effectiveness*, *readiness*, and *maintenance*. These measures are obtained essentially by monitoring three crucial, interrelated variables in the flow of information through the stages and procedures of strategic analysis: *data backlog*, *data relationality*, and *data signification*. This in turn creates the opportunity to monitor a number of other variables.

The matrix below shows the essential relationship of the measures to the three-stage model of strategic analysis developed in Chapter 2.

Model	Measures		
	Effectiveness	Readiness	Maintenance
Monitoring Threat Recognition Projection			

22 TOWARD EFFECTIVE STRATEGIC ANALYSIS

The remainder of the present section will discuss in detail the definition of the measures and the development of a measurement system. Figure 3 shows the area of emphasis in the present discussion with respect to the overall concept of computer-based strategic analysis. The area of emphasis appears clearly in the upper right corner of the diagram; the rest of the diagram has been partially obscured to indicate its comparative lack of relevance.

The Conceptual Framework

The development of measures is a most difficult challenge. Little previous research seems directly applicable. Yet the attempt to measure an elusive cognitive process of rationality and intuition requires building a conceptual framework. I shall sketch in this framework as background to discussion of the measures.

• **The Essential Dynamic of Analysis.** Analysis, a cognitive process, is described in terms such as "association," "interpretation," "meaning," "imagination," "intuition," "rationale," "induction," "deduction," etc. In analysis, essentially we associate different data and assign meaning. Because the strategic analyst makes projections, strategic analysis heavily involves the imagination.

A major point is the importance of the flow of information through the system ("throughput"). Given that input data is needed for the analytic dynamic to occur, and that strategic analysis takes place when the analyst reviews the data and assigns meaning, *this process occurs throughout the analytic stages as the data is throughput*. The basic dynamic can be depicted as follows:

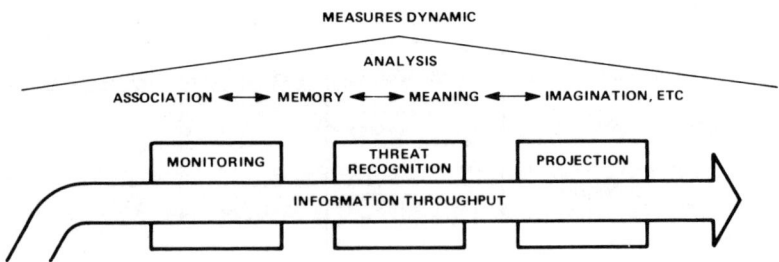

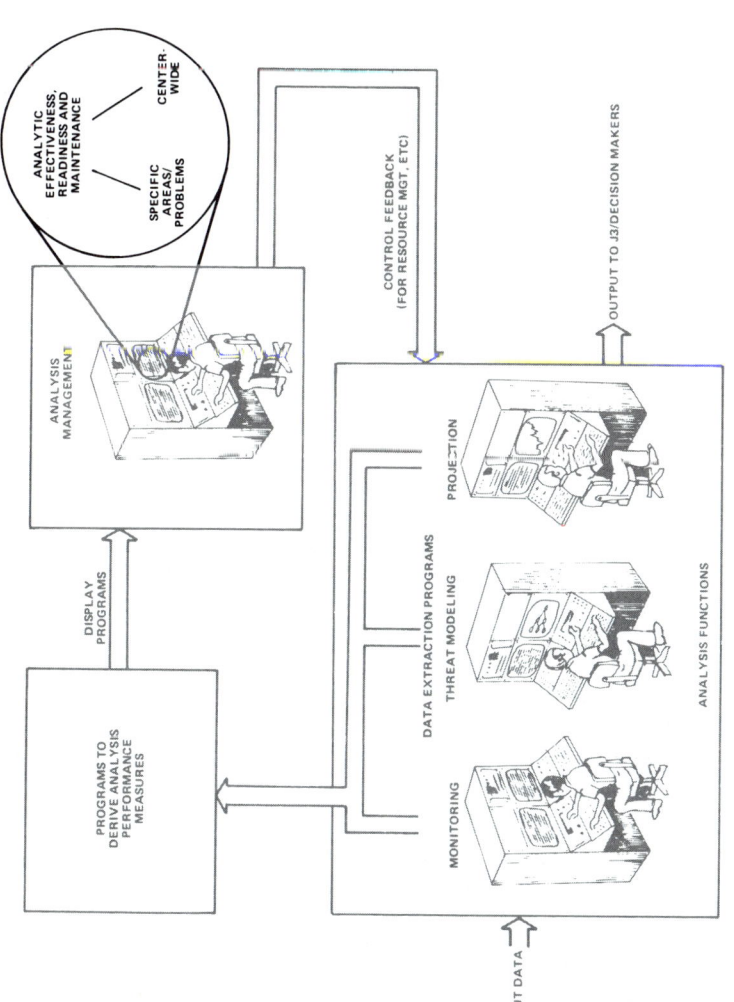

Figure 3. Measures Emphasis.

The essential measurement strategy is to identify key variables within the analytic functions in each of the three stages and then design a means of monitoring the variables.

- **Perspectives From Living Systems Theory.** Portions of James Grier Miller's work on the theory of living systems have been useful in developing a conceptual framework for measures development. Miller views living systems, ranging from cells to supranational organizations, as open systems composed of subsystems which process inputs, throughputs and outputs of various forms of matter, energy and information. He identifies six information subsystems as common across the range of living systems: the input transducer, the decoder, associative functions, memory, the decider and the encoder. The internal transducer is the sensory subsystem which receives markers from subsystems or components within the system bearing information about significant alterations in those subsystems or components. Association and memory are linked in representing the two basic elements of the learning process within a system. The decider makes judgments about the significance of information. The decoder translates input information into an internal or private code useful within the system. The encoder translates the final output into what might be designated a public code for dissemination.

Implicitly I have been viewing strategic analysis as cognition within a general system through which data progresses and whereby perspective is created. Figure 4 associates our analysis model with some of the structure of living systems theory. Such theory can aid us in deriving useful categories and structure for exploring the problem of analytic measures. In sum, strategic analysis, a process involving people, information, software, hardware and other elements, may be thought of as a system with various measurable states of activity, some appropriate and some to be avoided.

Figure 4 lists some key variables in information systems, some classic pathologies in such systems, and typical adjustments to the pathologies. In subsequent discussions of specific analytic measures, portions of Miller's living systems structure, particularly the variables, will be explored further.

- **Perspectives From Information Theory.** From information theory we must consider the problem of *meaning*, for meaning is central to the measurement of strategic analysis. We have developed the ability to measure rates of information as it passes through systems. It is the measurement of meaning that has eluded us, for meaning cannot be objectively quantified since it involves a human observer, a human participant, and

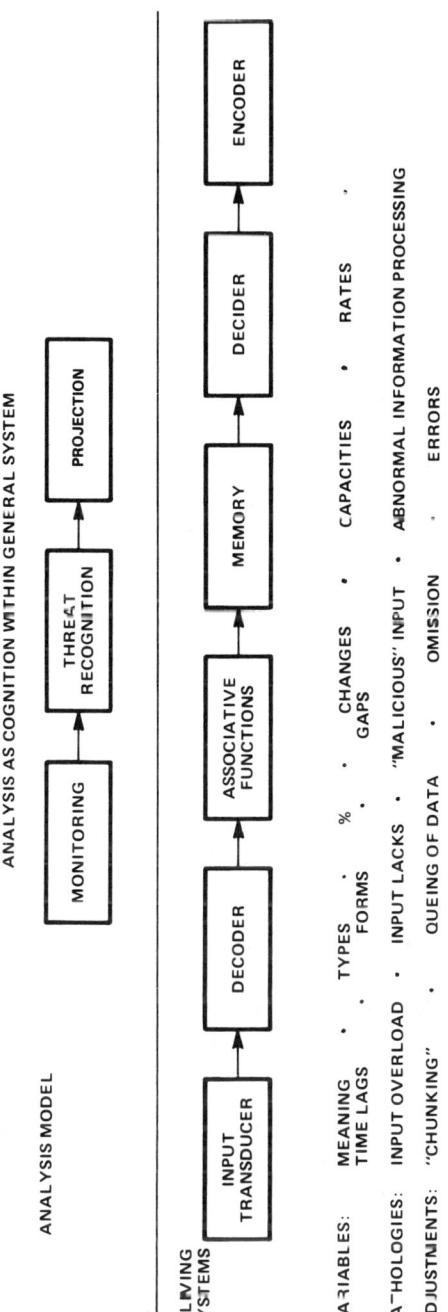

Figure 4. Strategic Analysis and Elements of Living Systems Structure.

hence subjectivity. Meaning may be defined as follows:

> Meaning represents the significance of information to the system which processes it. Meaning constitutes a change in that system's processes elicited by the information input, often resulting from associations made to it on previous experience with it. Furthermore, this change can occur immediately or later.

A crucial and related concept is *atrophy* of information. Information, taken as a retained model of reality, will suffer atrophy over time. In strategic analysis, for example, scenarios of possible threats obviously must be updated periodically. Models are static approximations of a reality that is always changing in manifold ways. The key concept for the measurement problem is that new information can counteract atrophy.

● **Perspectives From Learning Theory.** A fundamental issue is the relationship among input data, association, memory and imagination: the learning process of a strategic analyst as it entails mental operations on data. The discussion of analytic technique and methodology (Chapter 4) in part involves the learning process, especially as supported by cognitive technology — computer-based extrasomatic memories which represent collectively created, coherent records of analysis immune to obscuring processes of the human memory.

Basic Measures

As I have said, three basic measures of strategic analysis are: *effectiveness*, *readiness* and *maintenance*. Analytic *effectiveness* is an outcome, a result. Analytic *readiness* is a state, a condition. Analytic *maintenance* is an activity. The diagram below is meant to emphasize the interrelated, interdependent nature of the measures.

Figure 6 shows a simplified concept of a hierarchy of the measures and supporting variables. Effectiveness, readiness and maintenance (E, R, M) are derived by monitoring certain information-related variables which can be thought of in hierarchical terms. For example, data backlog at various stages in the analysis will become an important variable as will information-related "subvariables" such as percentages of information by type (for example, military, economic, political, etc.). Many combinations of variables might be monitored and analyzed. One imperative is to insure that enough data of interest is collected to allow multiple correlations among different variables in the future, including correlations whose interest was not anticipated earlier. As the remarks on living systems theory suggested, certain key variables are associated with information processing systems at all levels and should be accounted for in the design structure.

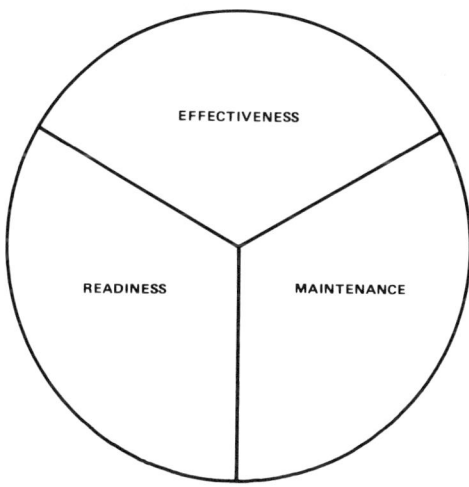

Figure 5. Basic Measures Diagram.

Figure 6 also identifies important variables which are not exclusively information-related, such as the dimensionality of the threat models and the use of antibias analysis procedures.

Finally, the diagram below conveys the design principle that all three measures should be monitored at each major stage of analysis. It is implicit that the measures must ultimately reach the analysis steps and procedures within each stage. Further, cumulative values must be obtained by integrating measurement data across and among the various stages to foster diagnosis.

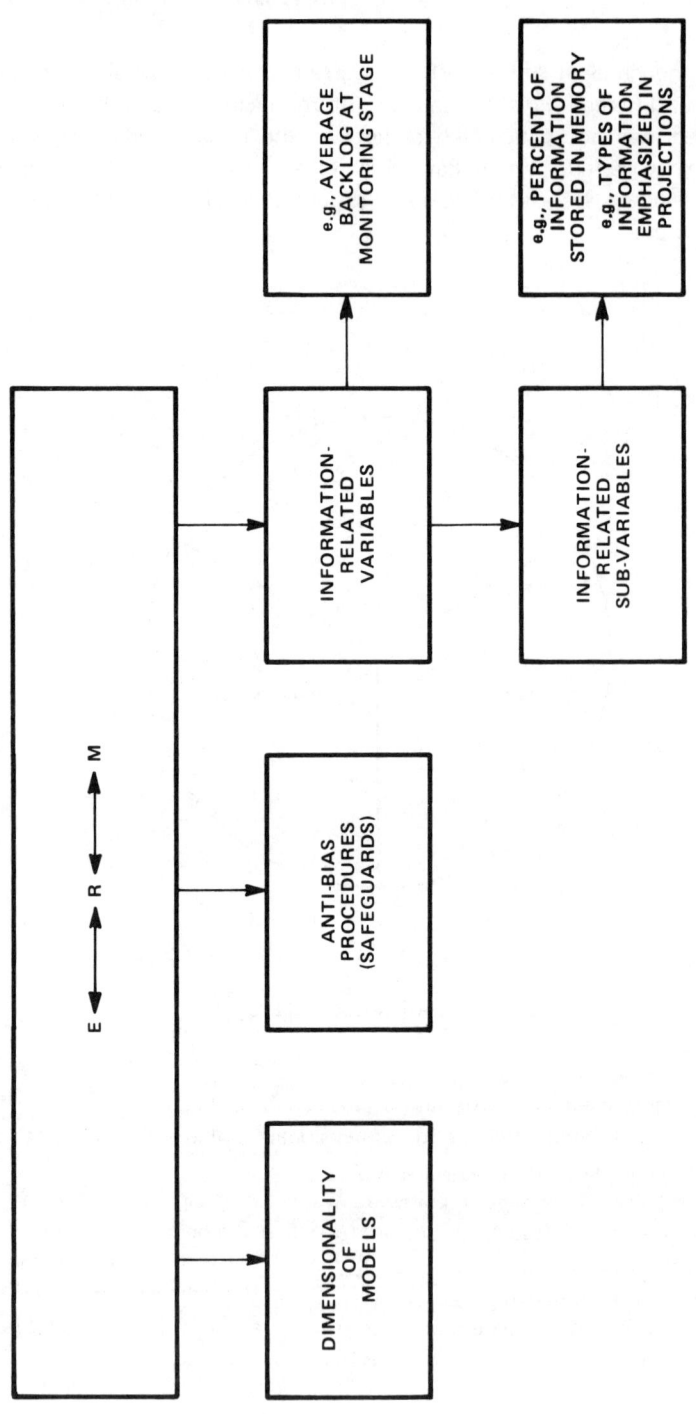

Figure 6. Hierarchy of Measures.

STRATEGIC ANALYSIS MEASURES 29

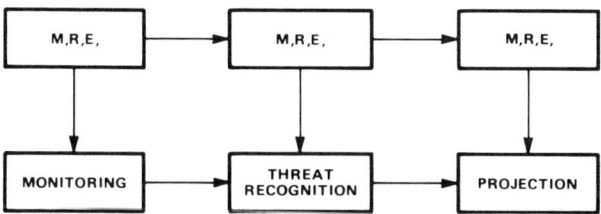

M,R,E ARE DERIVED FOR EACH STAGE/SUBSTAGE OF ANALYSIS; CUMULATIVE M,R,E ALSO DERIVED

These points have established a context for the detailed discussion of the development of measures to which I now turn.

Effectiveness

By far the most complex and difficult measure is effectiveness. Effectiveness is a measure of the ability to provide accurate, specified analysis in a timely manner. It differs from readiness and maintenance as follows: Readiness is a measure of the ability to carry out existing analytic procedures and methodology leading to effective analysis. Maintenance concerns the activity by which analytic procedures, models, and memories are sustained against atrophy, kept current and rendered operational.

The important point about the effectiveness measure is that there are really two kinds of effectiveness in strategic analysis: *estimated effectiveness* and *real effectiveness*. The strategic analyst confronts the future. *Real effectiveness* must be verified after the fact; validation is delayed, occurring through the *post mortem*. Yet there is a trap: hindsight analysis is perilous. There are sweeping arguments against the validity of historical reconstruction itself: under what conditions, for example, can a warning or alert be reliably connected causally to an outcome? Skeptics of *post mortems* on analytic failures (such as those which periodically occur in the intelligence field) cite problems in validly reconstructing past meanings of information to analysts and analytic communities, which is indeed a very critical problem we must deal with subsequently.

The diagram now shows the analysis measures expanded to include real effectiveness.

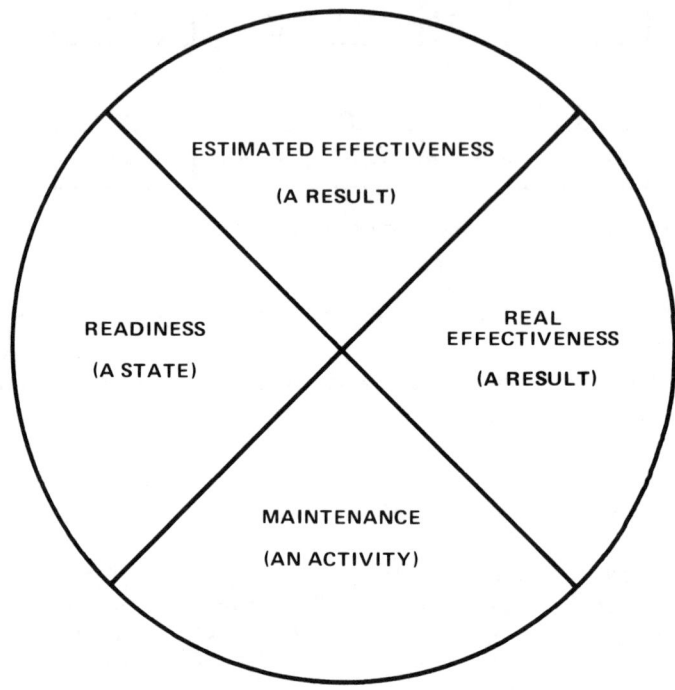

Estimated effectiveness, an *a priori* measure, is based on assumptions. These include the assumption that usually multidimensionality is good in strategic analysis: that it is better to take into account several frames of reference (political *and* economic *and* cultural, etc.) in interpreting a situation than only one. Another major assumption is that strategic analysis ought to incorporate safeguard procedures designed to offset the inevitable epistemological constraints and human tendencies toward bias which lead to susceptibility to deception, both natural and instigated by adversaries. *Estimated effectiveness* therefore becomes a wager that thoroughness, rigor and systematic approach in the methods of performing analysis will lead to greater real effectiveness.

Basic Obstacles to Effectiveness. The antithesis of successful strategic analysis is surprise. No matter what the context — political, military, corporate, etc. — effectiveness in strategic analysis must ultimately be measured in the context of avoiding surprise. I believe there is a shortfall in prescriptive research on mitigating surprise, although fascinating, useful discussions of such problems have been made in the field of military intelligence in such works as Wohlstetter's study of Pearl Harbor, Schlaim's study of warning failures in the Middle East, and Whaley's studies of deception. Of course, general discussions abound of basic epistemological and cognitive constraints on man's ability to analyze the past, present and future.

Certainly strategic analysis pursued energetically is a hard business, a struggle. The analyst might liken himself to Sisyphus, condemned forever to attempt to push a heavy rock up and over a hill, forever to be defeated near the top by insurmountable inertia. Basically, the analyst is implicitly striving for literal realism in a bewildering reality which does not necessarily arrange itself according to his mental constructs and images. Not that he has a choice in any final sense. He will do what, in Joan Didion's words, all humans do:

> We tell ourselves stories in order to live . . . we interpret what we see, select the most workable of the multiple choices. We live entirely . . . by the imposition of a narrative line upon disparate images, by the 'ideas' with which we have learned to freeze the shifting phantasmagoria which is our actual experience.

As I shall discuss later, there is some degree of kinship between the analyst and journalists and writers; between epistemological crises in the arts and fundamental problems in strategic analysis. There is less distance between concepts such as narratology and the problem of strategic analysis than might at first seem the case. The analyst, after all, seeks to interpret reality, to impose order on the flux of events. For example, the impulse to model possible threatening situations comes from the analyst's need to reduce a bewildering flux of events to some orderly system. Of late there is an announced crisis of confidence in epistemological terms on the part of some journalists, novelists and other artists, centered around issues of nonfiction versus fiction novels; the problem of interpretation in journalism; and other aspects in turn related to the perversity of reality in relation to our interpretive models and constructs. At the heart of the problem of analytic effectiveness is the problem of modeling a complex, multifaceted reality greater than our ability to perceive it. Our organizing models of reality can only be approximations.

Besides making sense of current situations (and often physically remote ones), the strategic analyst confronts and must explore the future. Of course very few events now are actually projectable by man with great specificity. Using the example of attempting to project the activity of a human being, Edward O. Wilson writes:

> Even though human behavior is enormously . . . complicated and variable . . . theoretically it can be specified. Genetic constraints and the restricted number of environments in which human beings can live limit the array of possible outcomes substantially. But only techniques beyond our present imagining could hope to achieve even the short-term prediction of the detailed behavior of an individual human being, and such an accomplishment might be beyond the capacity of any conceivable intelligence. There are hundreds or thousands of variables to consider, and minute degrees of imprecision in any one of them might easily be magnified to alter the action of part or all of the mind. . . . it may be a law of nature that no nervous system is capable of acquiring enough knowledge to significantly predict the future of any other intelligent system in detail.

Such problems in projection are compounded by the challenge of projecting changes in large-scale systems.

Given that strategic projections must be imperfect in the above sense, a major aspect of strategic analysis becomes more apparent. The strategic analyst will serve a valuable role by alerting decision makers sufficiently that they may in turn take action designed to preclude the possibility of an adverse situation and outcome. The distinction made above between unimpeded and influenced projections implies this strategy.

Given the severe constraints on strategic analysis, a natural approach to developing effectiveness measures is to begin with classification of the specific obstacles to analytic effectiveness, and then to conceive of measures which bear on capabilities and strategies for overcoming the obstacles.

Epistemological Obstacles. Several generic difficulties can be identified:
- **Limited Availability of Key Data.** Information is a model of the real world. By definition, information will be limited with respect

to total reality. Key data may not be available at a given time to the observer.
- **Analyst Limitations in Experience and Knowledge.** A given analyst's own knowledge and experience may not connect with given situations. In Soviet military literature, the example is given of a Japanese commander fighting on a remote Pacific island in World War II who refuses to believe reports of the atomic bombing of Hiroshima and Nagasaki. The Commander has no real concept of the development and potential destructive applications of atomic power in that period and is caught up in the local combat. He has no appropriate context or frame of reference.
- **Novel Situations.** A serious problem in strategic analysis, the novel situation seems novel because its characteristics have (or appear to have) little or no historical antecedents. Its features may appear innocent or go unnoticed for periods.
- **Similarity in Situations.** Some situations may be difficult to recognize because of their similarity in early stages to other possible situations. For example, when analysts develop a set of situation models, it is not unlikely that early events postulated in given models will be similar to those in other models.
- **Numerous Possible Situations.** Since there are so many variations of possible situations, it is always somewhat futile (if necessary) to attack the problem by assembling *a priori* catalogs of events presumed typical of significant conditions. The classic pattern recognition problem is a good deal simpler: a closed alphabet of patterns is assembled and the viewer must essentially sort out the patterns as they are presented in sequence. There are, however, effectively an infinite number of variations in real situations, so that an analyst must view preestablished situation models with appropriate skepticism.

Without pushing too far the distinction between "cognitive" and "epistemological," it is useful to discuss certain cognitive obstacles:
- **Restricted Perspective.** As Schlaim in particular discusses, analysts sometimes tend to work within narrow hypotheses. This is certainly understandable, for one can provide only so many models of the future. In both the arts and sciences, constructing such models (or scenarios) of real world situations is one of the most challenging and exhausting endeavors we can undertake. Further, once they are constructed there is a psychology of

preservation powerfully at work. One must also believe that the sense of imperfection about the models can at times diminish the will to struggle with them. In short, hypotheses may be hard won and defended stubbornly.

- **Overreliance on Theory.** Associated with restricted perspective, the problem of overreliance on theory arises when analysts apply theories of societal change based on previous strategic cases and fail to leave room for unprecedented causal factors in the current situation.
- **Overemphasis on Personality.** Analysts may attribute too much significance to "characteristic," "determining" personality traits of a foreign decision maker rather than considering that peculiar constraints in a given situation might cause the decision maker to act "uncharacteristically."
- **Bias.** A related problem is bias, which is seen in tendencies to establish low thresholds for incoming data that confirm prior assumptions and high thresholds for incoming data that may suggest other explanations.
- **Methodological Inflexibility.** Analysts may tend to work with only one or two "proven" techniques of analysis, rather than selecting from a variety of methodological alternatives in an effort to fit technique to problems. One basic assumption about effective analysis is that it is better to analyze information from several viewpoints, using a variety of techniques. The premise is that one of the analyst's defenses against the complexity of reality is to search for meaning by taking several analytic paths.
- **Susceptibility to Deception.** The problems of strategic analysis can be aggravated by the adversary as deceiver. Clever deception will try to exploit epistemological and cognitive weaknesses. The world itself deceives the human observer easily enough; clever deception operations can raise further the challenge to the strategic analyst.

Experimentation has, of course, demonstrated much about cognitive biases. Psychologists have charted limitations in man's memory, attention span, reasoning capability and other cognitive functions. These limitations considerably affect man's ability to process data to arrive at judgmental decisions, estimates, etc. As I have said, the limitations result in part from the complexity of our environment: it forces us to adopt simplifying strategies of perception, comprehension, inference and decision. These remarkably adaptive strategies allow us to deal with the multiplicity of information we take in and must process; yet ironically they result often in erroneous judgment. Thomas Belden, formerly Special Crisis Advisor to the Director of Central Intelligence, has denoted one consequence of cognitive biases in strategic analysis as the condition of "hardening of the categories."

Preliminary Framework for Effectiveness Measures. We can now perceive a preliminary framework for possible measures of effectiveness. A few concepts are:
- Ability to monitor on the basis of integrated indicators (economic *plus* military *plus* political).
- Existence and use of analytic safeguards to offset epistemological and cognitive problems in strategic analysis, such as:
  — Threat-search procedures for avoiding the trap of overlooking threats not previously imagined and modeled, e.g., techniques for learning from "ambiguous" and "incomplete" information.
  — Procedures/techniques for weighing opposing hypotheses.
  — Procedures that foster systematic use of a mixture of analytic approaches (e.g., various forecasting and prediction techniques) to increase objectivity and offset the inevitable bias resulting from overemphasis on any one approach or type of data.
  — Ability to perform easily the functions keyed to estimation and dissemination, such as: marshall evidence; develop statistical presentations; argue by precedent, individual indicators, and/or patterns of indicators; and employ modeling techniques.

Especially when we discuss information-related variables, we shall resume the exploration and development of concepts of measures of analytic effectiveness. It is useful now, however, to proceed to a discussion of the readiness measure.

Readiness

Readiness has been characterized as a state, a condition, a measure of the ability to undertake the procedural and methodological activity within strategic analysis. Readiness applies to all the stages, steps and procedures within the model of analysis; it must be measurable for any and all of them.

At the beginning we need to examine an important aspect of the relationship of readiness to effectiveness. We have discussed the problems of imperfect effectiveness in strategic analysis; that the sources of imperfection are epistemological and human cognitive problems. In an important sense, readiness is subordinate to, dependent upon, effectiveness. One can develop a system of analytic methods that entails certain procedures, techniques and methodology designed to foster analytic effectiveness, and upon which readiness can be based. The whole analysis system, however, must be imperfect in the face of reality. It cannot guarantee success.

The significant point is this: analysts in analytic groups must therefore be said to be in high (or low) states of readiness to follow an imperfect system of analysis. This relationship between effectiveness and readiness bears on a number of aspects of the present research. For example, consider the use to which measurement data might be put in a *post mortem* with respect to an "analysis failure." Given that analysts are working with imperfect techniques, the obligation to be effective must be defined within conditional terms; the obligation for readiness, however, may be viewed independently of the question of real effectiveness.

There is also an important relationship between readiness and maintenance. Readiness is a state which derives from maintenance activity, such as the activity of keeping the information in analysis models of threats up to date and thereby maintaining the currentness of the models.

Obstacles to Readiness. There are a number of obstacles, such as the following:

- **Atrophy of Analysis Procedures and Models.** Procedures and models tend to fall out-of-date. For example, a working threat model involving possible hostilities between the U.S. and another country will tend to go out-of-date because of evolutions in weapon systems, new political postures, economic changes, etc. The basic antidote is the process of analytic maintenance whereby new information is used to update the given model.

- **Difficulty in Implementing Analysis.** A serious obstacle to readiness can arise when analytic tools and procedures — including those up-to-date — cannot quickly and efficiently be employed by analysts. For example, quite realistic and useful threat scenarios may be developed spanning a wide range of possibilities, yet be documented in forms not easily used by analysts during time-critical periods. Computer-based support systems which include displays of models might greatly facilitate the ease of employment of the models and techniques. However, the accessibility of the computer-stored data then becomes a readiness concern.

STRATEGIC ANALYSIS MEASURES 37

- **Unavailability of Experienced Analysts.** An obvious obstacle to readiness is lack of training and experience applicable to different problem areas.

- **Analysts' Resistance to Analytic Procedure and Technique.** Analysts may become ill-disposed to employ certain procedures and techniques of analysis. Problems in psychological readiness can arise from a variety of causes and are discussed in succeeding chapters.

Typical questions which center on the issue of readiness are: Is there a backlog of data to filter through given models? How long does it take to retrieve various portions of memory stored in given computers? How long does it take an analyst to compare past and present data? How comprehensively can a given number of analysts worry different sides of a problem within a limited time frame?

Preliminary Concepts of Readiness Measures. Clearly there are several dimensions to readiness within the strategic analysis problem, such as: *analyst readiness*, *system readiness*, and *supporting technology readiness*. Analysts must have experience and training to assure analytic readiness. The system of analysis itself — the models, techniques, procedures — may be more or less ready. Further, the operations the analyst performs in recalling information, displaying information, matching disparate data sets, etc., involve readiness.

Readiness as a state, a condition, is ultimately the state of the entire analysis system. Readiness must be reflected in a set of values on some scale which the key variables of that system have at any given instant. Subsequent discussion of information-related variables will allow us to explore further and define detailed readiness measures.

Maintenance

The third major measure is analytic maintenance. Maintenance is an activity; it is a measure of the sustaining of analytic procedures, the keeping current of the analytic memories. Maintenance is, in short, partly accomplished through the act of analysis itself.

It does not matter that we are talking about a cognitive activity: when analysts update models, refine procedures, review incoming data, and perform other analysis functions, they are maintaining readiness, readiness of their own capabilities and readiness of the supporting

38 TOWARD EFFECTIVE STRATEGIC ANALYSIS

methodological and technological systems. This in turn influences effectiveness.

Obstacles to Maintenance. Some examples are:
- **Failure to route incoming data through analytic procedures, techniques and models:** When new data is not routed through the analytic process, atrophy is risked. Maintenance arises out of the connection of data throughput to the cognitive functions of analysis. If the analyst does not periodically run the data through the system, then the system must lack maintenance.
- **Failure to conduct adequate checkout of technical support systems:** This classic aspect of maintenance refers to preventive maintenance of hardware and software support systems. A display system that frequently is down for repair, an unreliable computer-based system, and a software package not fully debugged are obvious examples.

Preliminary Concepts of Maintenance Measures. Typical maintenance questions include: What is the periodicity with which new data has been run through the system of analysis? How much preventive maintenance has been done, and when, with respect to given support systems?

As the diagram below shows, we have now completed a general discussion of the four major measures of analysis performance.

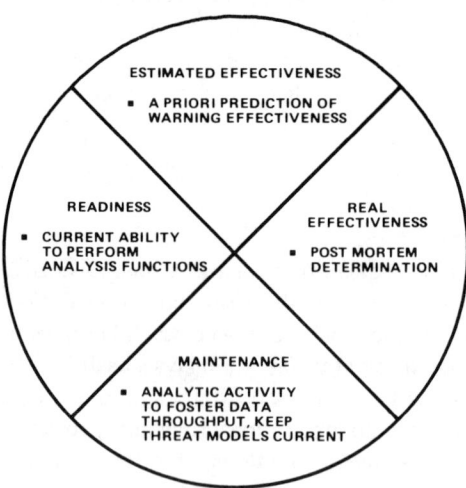

Key Information-Related Variables

To derive real and estimated effectiveness, readiness and maintenance, key information-related variables must be monitored. The variables should have certain characteristics. They should be interrelated within the overall process of analysis; they should be recognizable (monitorable) and intelligible to observers such as managers and analysts controlling the system; they must be sensitive and significant for diagnostic purposes; and their change must be measurable.

Three information-related variables have been developed as fundamental to deriving the measurements:

- Data backlog
- Data relationality
- Data signification

In the discussion which follows, each of the variables will be defined, explained as to choice, and exemplified. Specific connections between the variables and the measures will then be developed. The basic intent is to show how the key variables combine with the measures to produce a meaningful, tightly coherent system of linked, interrelated measures.

I will also explore certain important subvariables which should be examined on the basis of empirical data generated by monitoring the three key information-related variables. That discussion will be followed by a consideration of the best uses of the data, with primary focus on the problems in strategic analysis that might best be approached with various kinds of data generated by the measurement system.

Backlog

The classic variable of *backlog* essentially involves identifying important information anywhere in the analytic process which is in a waiting position, which has yet to be analyzed by strategic analysts; which has, in short, not completed throughput. We must think of the backlog in relation to all points — all stages, procedures and steps — in the total model of the analytic process.

It is desirable to measure a number of backlog values, such as: the size of the backlog, the nature of the backlog (e.g., the types of data which comprise it), the duration of the backlog, and such dynamic measurements as the rate of decrease or increase of backlog, the acceleration values within the rates of increase and decrease, and other measures.

These of course could be related to other variables such as number of analysts, technical support, etc.

Backlog is a fundamental variable within the system, for a great portion of the entire approach is founded on the need and desirability of throughput. Readiness, maintenance and effectiveness have all been related in different fashions to the process of throughput. (In my extended visits to strategic analysis centers, I have become fully aware of the relatively large volumes of input data typically on hand that must be processed and analyzed. I will discuss the "volume" problem in a later chapter, indicating for now that I do not view it as an insurmountable obstacle to achieving necessary backlog determination.)

Relationality

The *relationality* variable involves the extent to which throughput data is related to the full system of analytic stages, steps, procedures and methods in the overall methodology. Relationality embodies the following question: was the data analyzed, or is the data now being analyzed, through all or some (and if so, which) of the procedural routines, models and projection techniques which comprise the system of analytic art? The analytic system is the analyst's best (if inevitably imperfect) offense against the epistemological and cognitive problems in strategic analysis. The extent to which the analyst appropriately follows that system bears directly on, first, estimated effectiveness; second, real effectiveness (to the extent that the system has been demonstrated as empirically effective); third, maintenance, since the act of analysis itself comprises a large portion of analytic maintenance; and finally, readiness, since readiness is the state or condition of certain analytic abilities. In short, relationality, like backlog, is fundamental to the basic measures.

Furthermore, relationality and backlog have an important complementarity: reduction of data backlog *should* occur through relationality to the procedures and methods within the analytic system. The two variables are interrelated; together they indicate far more about strategic analysis than taken separately.

Signification

The most important of the three variables, *signification*, is also the most difficult, the most experimental. Signification involves the meaning assigned to data by analysts. Signification is captured by recording changes made by analysts at certain nodes in the analytic process; for example, changes in variables *in* analytic models and changes *to* the structure of these models and the reasons for the changes. As discussed above, meaning, taken as constituting the significance of information to a system which processes it, incurs a change in that information system's processes elicited by the information input, the change often resulting from associations made to the data based on previous experience with it. Change can occur immediately or later. Given that in part we are viewing analysis in systemic terms — a process with stages, steps, procedures — the monitoring of meaning through changes expressed at nodes in our functional model becomes a natural strategy.

The strategic analyst essentially attempts to determine the meaning of incoming information in terms of the degree (if any) to which it may signify the emergence of a crucial or threatening situation. The Soviets distinguish this act as a form of decision making, referring to it as an *information decision*. (The analyst is also interested in recognizing nonthreatening situations and therefore assigns meaning to the data in any event.) The crucial point is this: if we wish to explore real effectiveness, we must be able to perform *post mortems* which essentially are reconstructions of past analytic states and perspectives, which leads us directly to the need for a selective history of meaning.

The question invariably arises: Why take on a problem as complex and subtle as the problem of measuring meaning? We know that objective quantification is not possible at the present time. We know that contemporary models of human cognition are by no means complete, and that we can make no pretense of fully understanding and hence capturing the *detailed* (and mercurial) processes by which meaning occurs in a living system. What, then, are the prospects for achieving useful results? What are the circumscribed goals? To answer these questions, we must explore further a central issue in strategic analysis.

The Problem of Real Effectiveness. Real effectiveness is, of course, the ultimate frame of reference for strategic analysis. Its measurement vehicle is the *post mortem. Post mortems* can be formal, informal, limited to "failures," conducted in all seasons, etc. But stated simply, real effectiveness involves these questions: Looking back, how well did we do? And why?

Besides the political and technical difficulties, the *post mortem* process is beset with evidentiary difficulties arising from cognitive and other constraints. These latter kinds of difficulties in conducting hindsight analysis will always (and fundamentally) complicate the determination of real effectiveness.

A useful starting point is this: analytic *memory* is crucial for *post mortems* and the determination of real effectiveness. But there are serious constraints on analytic memory. The problems may be expressed usefully in terms of cognitive biases such as discussed earlier. Richards J. Heuer, Jr., has recently reviewed and constructively interpreted the research literature in this area. He concludes that:

- Analysts tend to overestimate the worth and accuracy of their past analytic judgments.
- The consumer of the information (in this case the user of the analysis product) tends to underestimate the value of that product to him.
- The manager conducting a *post mortem* tends to conclude that events were more readily foreseeable (or less so) than was in fact the case.
- *In short, we seem to have a systematic tendency toward faulty memory of our past estimates.*

The explanation for these problems comes largely out of experimental research. As Heuer has told us, there are several key points:

- It appears that additional information available for hindsight analysis actually changes our perceptions of a situation so naturally and so immediately that we are largely unaware of the change.
- The new input data is immediately (and probably unconsciously) assimilated into our prior knowledge.
- If the new data adds significantly to our knowledge — tells us a situation outcome — our mental images are reconstructed.
- Apparently it is then virtually impossible for us to reconstruct our prior mental set.
- Thus, we may recall *but not reconstruct.*

Ironically the human being at this stage of his mental development has a memory, a data base, which is perturbed, altered and, in fact, progressively lost in important ways; and lost, moreover, somewhere within the vital and necessary act of taking in new data, throughputting the data within the human cognitive process, and changing perspective on the external world. Thus in the act of learning we also forget.

In principle, there is a remedy: extrasomatic memory. In discussing the development of, and outlook for, human intelligence, Carl Sagan and others have stressed the importance of information stored outside the body — the early invention of libraries of the written word, and the future prospects of more advanced, interactively dynamic forms of extrasomatic information bases which use computer technology. In some commentaries, there is a sense of urgency, the belief that the world is changing more quickly than ever before, and that to perceive the changes and react in ways that foster survival, learning systems supported by advanced cognitive technology are crucial.

The image of the world changing ever more quickly, with the change substantially of our own making, and the change increasingly taxing our perceptual, cognitive and hence survival powers, obviously and fundamentally bears on the strategic analysis problem. For the present, it is instructive to cite only one of the many problems in the current dilemma. This is the problem of the increasingly stringent requirements to monitor significant change, especially threatening change, in areas physically remote from the immediate environment of the perceiver. The "remote" quality obviously carries over into political, cultural, economic and other dimensions, thereby complicating recognition by the watcher on all levels of analysis. The need to watch and analyze such events is not, of course, new; the problem is that the reaction times and other constraints are growing more severe. Yet the evolution of human intelligence may be at a stage in which we are still operating in part as if the threat from extinct predators is more real than the complex international/intercontinental threats of today. Can there be any doubt that unaided human perception, analysis and decision making are inadequate to the challenge? It seems obvious that cognitive technology is essential to survival. Certainly it is my view that such technology is now, and increasingly will be, vital to the strategic analyst.

Focusing on the essential issues of strategic analysis — the problems of real effectiveness, *post mortems* and learning — let us suppose that as analysis progressed in an analytic community we could succeed in steadily, continuously capturing signification values (as well as backlog and relationality values) and storing them in an extrasomatic memory, where computer programs would perform search-recall-reduction-reconstruction functions and display the results in various forms, including boiled down sequential or narrative accounts of the history of the development of given analytic rationales — the input data, its impact, the assumptions and the methodologies

There would be several benefits. The recorded memory of meaning would be preserved safe from the perturbations in the human memory. Further, the basis would be laid for an improved institutionalized procedure for reliably obtaining a permanent, accurate record of certain key analytic constructs — constructs developed by changing communities and groups of analysts — which we would need for *post mortems* and the measurement of real effectiveness as well as for operations research and the definition of R&D requirements. Among other things, the data base developed would be beneficial to research and innovation in methodology, analytic incentives, and in other critical areas in which the basic constraints on analytic effectiveness must be challenged if there is to be progress. Finally, we would be creating the basis for continuous *post mortems* internal to the analytic community and conducted by analysts and managers at all levels. Handled properly, this capability would hold some advantages over (if not necessarily remove the need for) *post mortems* done essentially by "outsiders" such as appointed committees who explore an analysis failure. Of course, the computer-based technology supporting the extrasomatic memory must be designed such as to eliminate unacceptable drain on analysts' time. Obviously behavioral ironies created by such a system among analysts and managers is also a serious potential problem to weigh carefully.

Levels of Signification. Two related questions now arise: To what extent can we measure signification? To what extent should we measure it? The answers to both involve our knowledge of human cognitive processes. As we know, brain research has led to mappings and models of the brain at various levels and dimensions — chemical, electrical, neurological, functional, etc. Mental processes have been classified (intuition, rationality) and associated with cognitive activities (counting, writing, language comprehension, nonverbal ideation, etc.). More refined models are being developed as new research leads to further understanding. We have known for some time, however, that the operations of the mind on data are enormously complex. There are, for example, millions of light receptors associated with the retina of the human eye which receives visual input. There are perhaps a million ganglion cells which receive signals from the receptors in the retina for processing prior to the throughput of the data to a section of the cerebral cortex, after which other microoperations of great complexity occur. Thus when we refer to the information subsystem within a human system (input transducer, decoder, encoder, associative functions, memory, etc.) we are considering a process which at its fundamental levels involves millions of parts and processes.

The need therefore arises for a tractable yet viable model of human cognitive activity, constructed at a level of detail suitable for present purposes. Too simplistic a model would not yield enough variables for productive strategies of cognitive technology to emerge. Too intricate a model — a detailed neurological model, for example — would cloud the focus on key strategic analysis problems.

The attempt to develop a functional model of the cognitive steps in strategic analysis (Chapter 2) is judged an appropriate general strategy. However, we must search for a real and usable analog in human cognition to the potential signification-measurement points in the model. Recall, we have adopted the following strategy for measuring signification: we will create a set of models of threat situations; projection models; and other nodes in the analytic process which are (a) subject to change by an analyst as a function of the significance of input data, (b) visible to monitoring, and (c) likely to reflect important changes. Recorded changes would comprise the memory, which should be extrasomatic. (We will exemplify and discuss the design features of such nodes in detail in Chapter 4.)

We know that we cannot claim that such an approach will capture *all* aspects of signification. This is because of factors such as (a) we do not now understand mental processes thoroughly enough, (b) changes in the analyst's mind caused by input data and reflecting meaning are "subjective," (c) the changes are virtually continuous, (d) they are perhaps not fully conscious, and (e) they are always subject to loss in the sense discussed above. Thus we will have an *incomplete* record of signification. For example, even within a system in which analysts indicated and recorded signification according to certain criteria and within a schedule, there would still be some loss, especially in terms of the items listed immediately above.

Can we, though, have a realistic and useful if limited record? One that serves both a valuable present purpose and becomes the first-generation memory in an evolving series of increasingly refined and productive memories? In the author's judgment the answer is, Yes.

First, there is an apparent solution to the problem of finding a functional analog in the human mind to the extrasomatic system of measuring signification. Specifically, there is something of an analog between an extrasomatic model of a threat situation stored in a computer and recallable in display graphics form and whose nodes of activity can be changed by an analyst through the new meaning conveyed by input data (signification); and configurations carried within the human brain which cognitive psychologists call *schemata* or *plans* (sometimes described also as "templates," "scripts" or "knowledge structures"). One technical definition of a schema is:

> ... a configuration within the brain, either inborn or learned, against which the input of the nerve cells is compared. The matching of the real and expected patterns can have one or the other of several effects. The schema can contribute to a person's mental 'set,' the screening out of certain details in favor of others, so that the conscious mind perceives a certain part of the environment more vividly than others and is likely to favor one kind of decision over another. It can fill in details that are missing from the actual sensory input and create a pattern in the mind that is not entirely present in reality. In this way the gestalt of objects — the impression they give of being a square, a face, a tree, or whatever — is aided by the taxonomic power of the schemata.

The schema plays a decisive role in human cognition, particularly with respect to such analytic functions as interpretation, recognition and extrapolation, and hence seems an appropriate element within human cognitive operations with which to draw a correspondence in designing a functioning extrasomatic memory of signification. From the imputed correspondence between the schema and the extrasomatic memory of meaning, configured as it is in models which are themselves a kind of artificial schemata, we develop a possible basis for a realistic and useful record of analytic meaning.

One major prospect is that the imputed correspondence offers hope not only of at least some initial realism in the structural design of the extrasomatic memory, and hope not only that the analyst has no fundamental behavioral aversion to the structure (that the "human factors" issue is not overriding): it also offers hope that the extrasomatic memory will quite naturally improve over time, be refined, made increasingly more compatible with human usage, and hence become more powerfully reflective of signification, since one might reasonably expect that continued interaction between the analyst and the extrasomatic memory (in practical terms, of course, a man-machine interaction) will naturally lead to a convergence, a humanizing of the extrasomatic memory.

There remains, however, the question of incomplete signification. Ordinarily we will not be able to record signification operations in the brain on a microsecond-to-microsecond basis. Ideally should we monitor signification at such levels? Since we cannot, what is the likelihood that we can design useful signification measures? In my judgment we will very probably lose a great deal, but even now not enough to impede important measurements. Moreover, a grasp of the microprocesses of signification may not lead to rapid improvements. Assuming that we someday understand them, these microprocesses may simply reveal considerably more detail about an important fact of which we already are only too well aware: human fallibility. We can now estimate fairly accurately storage and processing capacities of the mind. Without knowing fully its cognitive microprocesses, we can also recognize beyond any doubt that the progress of the human mind in understanding reality has been painfully slow. We do not need to know more at this point in order to pursue signification usefully.

In later chapters concerning system design, key nodes are identified in the analytic process believed important in capturing signification, and the design of these nodes is exemplified in some detail. A simple example is certain types of information filters existing in threat models by which input information may be associated with preestablished meaning. In this type of signification node, input data may be clustered around certain nodes which model events hypothesized likely given the activity of certain indicators.

Obviously it will be necessary to collect and review empirical data to assess the utility of specific means of measuring signification in an analytic process.

Before continuing the discussion of signification, it is useful now to attempt to envision more completely the analytic process we are seeking to measure. For in exploring signification, we have begun to raise very complex and subtle issues of cognition and inference, issues which are at the heart of the challenge of measuring analytic effectiveness.

A Symbolic Representation of Strategic Analysis

At this point, let us reconsider the process of strategic analysis. It served our purposes earlier to represent that process functionally; but in developing performance measures we must go further and form a sharp *central image* of strategic analysis. As best we can, we must develop a sense of the constraints: those human and other causes of analytic difficulties cataloged above. More importantly, we must begin to imagine new prospects for strategic analysis. Above all, we must *see* these things.

There is also another need for a central image. Unfortunately we are very far from a detailed and scientific understanding of many of the challenges against which analytic success would be measured. Traditionally we have represented some of the challenges in philosophical, mythical and even mystical terms. We must continue to do so. This can hardly be surprising since many problems of strategic analysis reflect processes of human cognition bound up with profound mysteries likely for many more centuries to puzzle man in his quest for greater self-knowledge and understanding of reality. With no apology, therefore, I pursue briefly an attempt to represent in one image the essential process of strategic analysis — its elements and dynamics — in what can only be a sentimental, imprecise and largely symbolic fashion. The justification is the hope of sponsoring a working sense of the human analytic process, its constraints, challenges and potential. For it is this process which we seek to measure.

Discussion of Image. Figure 7 shows an artist's rendition of the strategic analyst at work. The figure is simply one way of imagining and configuring an endeavor whose ideal state we cannot now fully imagine. The drawing is intended as a coherent image of the meaning and substance of strategic analysis as we now pretend to understand it. We shall examine the image in a general fashion, returning later to concentrate on various portions as we further explore certain processes represented there.

Note first the human analyst. He is operating a computer-based support system by entering commands through a keyboard and monitoring a display showing his models and representations of reality. Our concept of modern strategic analysis presupposes a crucial role for information technology. It is through the computer, the electronic visual medium of automated displays, and unimagined future technology in support of cognition, that some of the major problems of analysis may be mitigated.

In the great length of known and surmised history, the human analyst is a very recent creature whose near ancestors — those involved in later evolutions of intelligence — are indicated by the heads of man's predecessors. It seems essential that the image instill a sense of both the longevity and the brevity of human intelligence — the length of its whole development, the shortness of its stages. And the image should suggest the increasing incongruence between painfully evolved properties of intelligence needed for earlier survival and the all-too-sudden strategic imperative, the growing necessity of searching ahead of time — indeed, outracing time in a much deeper kind of vision, an unprecedented form of reconnaissance — thereby perhaps allowing pursuit of a safe course through increasingly dangerous times. In the broadest terms, the strategic analyst searches the future for signs (really, "conditions") of his perennial enemies, of such apocalyptic disasters as war, economic catastrophe, famine and disease.

Nothing seems gained by presuming to question the character and timing of the evolving design of human intelligence — by wondering about what we have lately perceived to be its flaws which might imperil survival. Not only is it irrelevant to complain, but it deepens those ironies in nature which mock our vanity. A more practical perspective is simply that with information science and technology we are making tools for bringing order and mastery, and hence for survival.

In looking again at the analyst, we see him involved in several processes. First, and always imperfectly, he seeks to imagine and model future states of reality, symbolized by the carpentered model of the world shown beyond the dashed line marking the domain of the future. Question marks, signs of riddles, signify hidden, elusive meaning and stand as man-made symbols of the interpretive impulse. The carpentered model of the future — the analytic art, the output — is built to diminish, if not cover, perennial questions.

A second major process is receipt of data input from reality: from the analyst's immediate environment, for example, the imagery on the face of the computer-driven display; or indirectly through reports of events not observed by the analyst. The analyst "forms images" such as scenes, objects, personalities, etc. In part, he utilizes schemata to comprehend the nature of the input. He reviews the input information — looks for associations — on the basis of information both in his own memory and in the machine memory. He analyzes that data through models and algorithms, some operable through the machine programs.

A third major process, depicted in the enclosed area beginning immediately behind the analyst's head, occurs when the analyst stores in his own memory the images, models, and other constructs developed through his sense impressions of reality. As memory reaches further into the past — symbolized by the backward flow of time on the clocks, the clocks themselves regressing through older designs — the obscuring processes of human memory discussed above are symbolized by the increasingly curtained stages, their scenes and sets progressively less visible.

The circularly enclosed extension of the analyst's memory represents that very real part of humans called the "subconscious" or "collective unconscious" or "genetic script" or various other names. I have no interest here in reviewing the tangled philosophical and scientific controversies underlying such names and concepts. The figures within the circle are all familiar mythical and common symbols of what traditionally are taken as dominant forces within man: images of evil drawn from memories of prehistoric reptilian enemies; of violence and man's fashioning of weapons; of the basic human orientation toward parents, love and sex; of the search

Past

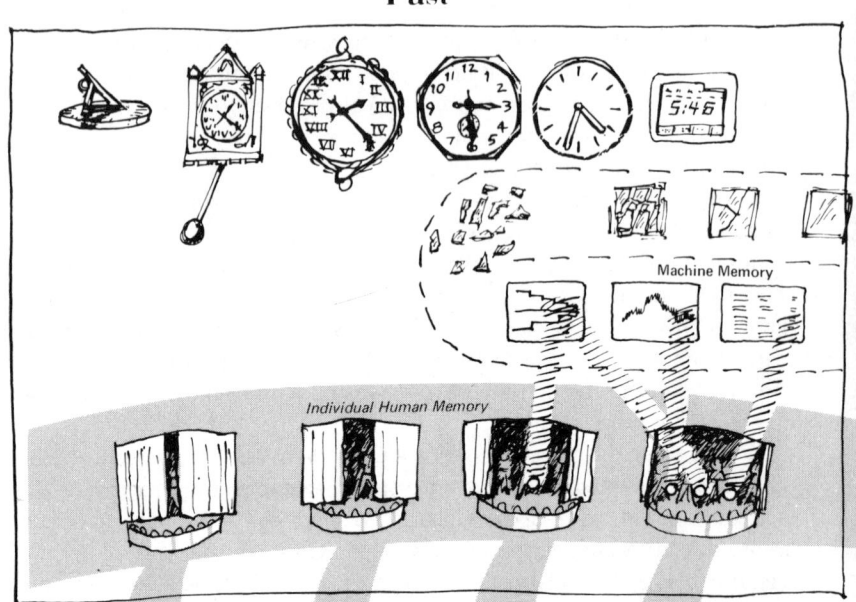

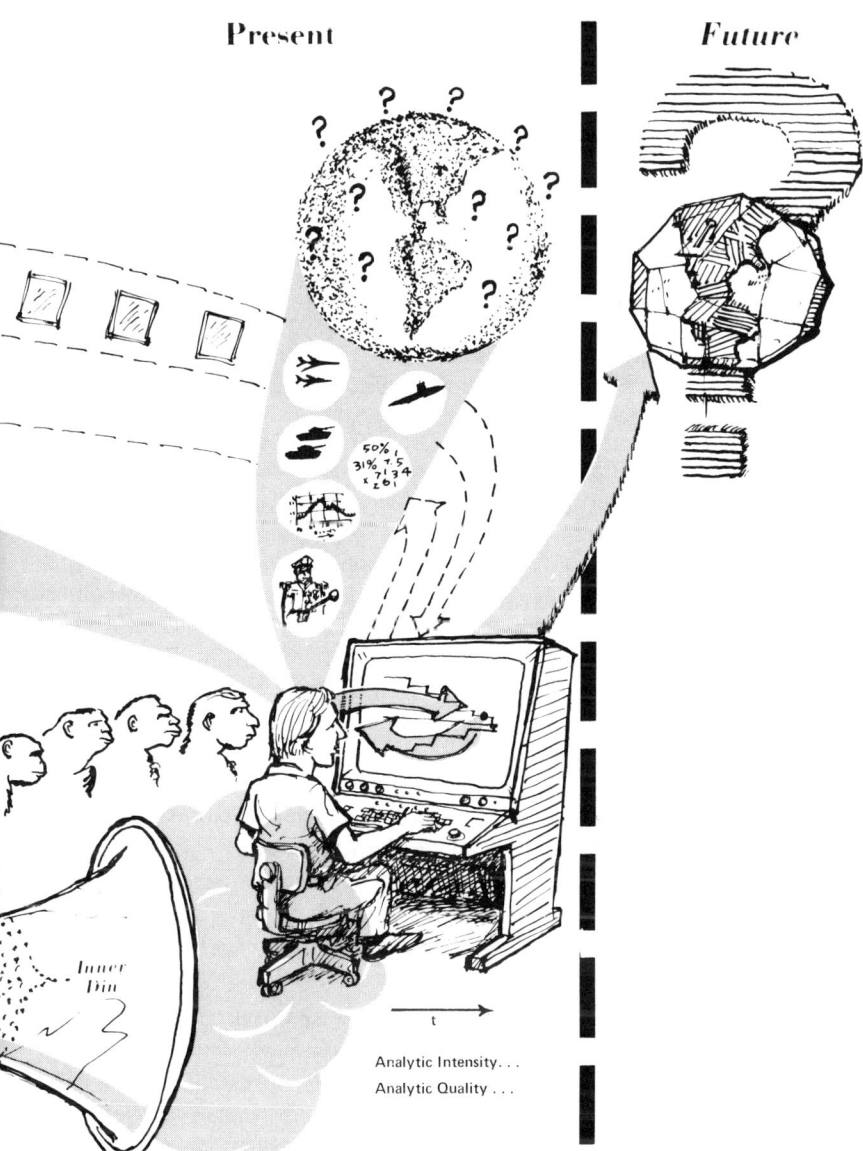

Figure 7. A Symbolic Representation of Strategic Analysis.

for food and for spiritual and aesthetic fulfillment; of the striving for order; and of the imperative for freedom.

Suffice to say that the strategic analyst is affected by all these factors and more. The analyst must strive against his own nature: physical and mental exhaustion; lapses in concentration; mind set; inadequate brainpower to understand the meaning of the information he has; physical constraints on his ability to process information; constraints on his memory which retard learning; and old fears and drives bound up with his survival yet distracting him from the strategic perspective. It is a constant condition in the form of an Inner Din distracting him from sustained cognitive efforts. Inefficiency, interruption, bewilderment, rationalization — such problems mean that the analyst has really very little peak time at a given stretch.

Given these limitations, it is important to consider the machine memory, symbolized by the arrow which ends at the top of the computer and then traces in a circular pattern back and over the analyst before returning to the computer. The arrow contains several mirrors and the fragments of a broken one. The mirrors symbolize the *artificial schemata* considered above in the discussion of signification (and which will be explored in detail in Chapter 4). I have a certain sympathy with the critical reviews by Dreyfus and others of the idea of "artificial intelligence": can a machine really perform intelligently in a human sense? Through the notion of artificial schemata, I am saying essentially that the machine can store, display and modify analysts' models or representations of states of reality; and that these models, through their formal designs and through the changes made to them when new meanings are assigned, also serve as limited reflectors (and subsequently records) of some portions of the thought processes in analysis. Imperfectly, these mirrors can capture for future analysis, as well as for measurement, such analytic processes, and preserve them safe from the obscuring processes in human memories.

We must evolve more and more refined artificial schemata, discarding earlier versions as we learn; hence the symbol of the shards of the shattered mirror. Yet we may not soon design schemata capable of reflecting other than very small portions of the stream of analysis. Clearly, we face a crucial long-term challenge in the art of constructing models, artificial schemata, which must hold within their designs techniques and processes of analysis to isolate and preserve decisive portions of cognition, giving us vital evidence for discovering needed refinements in analytic art and opportunities for more productive technology; as well as fostering analysis in general by promoting continuity in thought. (Regardless of their form, it is also implicit, of course, that the models act as filters, that they "force" or "shape" analysis to some extent; therein lies both a strength and a weakness of models, a subject to be discussed in Chapter 4.)

Let us now consider the actual analysis performed by the analyst. This analysis is symbolized by the arrow repeatedly interfacing with the computer display as analytic iterations occur. Fundamental to the performance measures is this concept: the machine not only preserves some portion of meaning and the analysis that led to the assignment of meaning; the machine, through its speed of operation and the rapidity of comprehension made possible by its sophisticated graphics, also leads to more *intensified* analysis. The quantity and quality of analysis over a given period increases. More iterations across models, involving more input data, are made possible. Just as a composer at a piano translates his ideas into sound directly through a keyboard controlling sound-producing mechanisms, and from there perhaps into a recorder, so ideally the analyst should have a computer-based support system that directly accepts, records and displays pertinent analytic functions and results at due speeds. Such capabilities are vital in promoting the analytic heuristic. They are also fundamental to the formal concept of analytic *effectiveness*. The fact that the aesthetic of the fine arts and the aesthetic of strategic analysis are different — that should the strategic analyst produce analysis for its own sake he may subvert his relationship with the decision maker — does not invalidate the basic analogy.

It is crucial to understand that computer/display technology should be harnessed to allow analysts to operate with extrasomatic support systems designed to enhance both the cognitive processes of analysis and the measurement of analysis. Analysts must be aided in the creation of extrasomatic memories in the form of models or schemata; and to manipulate them readily through interaction with the machine for purposes of changing the structure of the models, recording analysis processes and results for measurement, and recalling the analytic memory. Consider the fact that the memory structure I shall outline in Chapter 4 — a substantial array of models and procedures complementing the stages and steps in the functional model of strategic analysis — is coherent, large in scope, thorough, and yet not readily recalled, let alone manipulated, by the unaided human mind: the machine support is absolutely essential. Yet with the machine and with quite rudimentary training, the entire structure becomes readily usable in all dimensions — analysis, measurement and management.

We should also note the display which the analyst is observing. One of the obviously essential applications in displays is the use of color, animation and other cues to highlight changes, lend emphasis to portions of analysis, isolate rationale, show threads of continuity in analytic evolutions, indicate varying levels of specificity, assist in memory recall operations, single out unresolved input data, and many other tasks. There is every prospect that color, animation, flashing cues and other devices that lend themselves to electronic displays will form part of the rhetoric, the style, of the electronic visual tradition. As we review the analysis output formats of models and maps of cognition in the present approach (shown in Chapter 4) and imagine future refinements, it becomes clear that such visual devices will become extremely valuable.

Among the conceptual foundations of analytic effectiveness, then, is the assumption that more analysis concentrated and intensified over shorter periods — those times over which the human analyst can in fact sustain analysis — will improve the capabilities of the strategic analyst; that such intensified analysis will enhance the analytic adventure, building motivation for interpretation, sleuthing and the pursuit of understanding in the face of uncertainty. Implicit here is an antidote to the extremely important and subtle problem of bias. Conceivably, intensified analysis should make the analyst less likely to depend on single hypotheses, to succumb quite as readily to mind set, wishful thinking and other common pitfalls. The ability rapidly to create and compare hypotheses, challenging and changing them, should foster cognitive fitness — mental toughness — vital to strategic analysis. It should lead to more relativistic analytic postures and inspire the critical thoroughness and devil's advocacy vital for bettering strategic analysis. In short, improved quantity, quality and productivity of analysis presumably will enable a strong attack on the basic epistemological and cognitive challenges in strategic analysis.

Transition

We were considering signification, its problems and prospects, when I digressed to enlarge the concept of strategic analysis. We may now resume the exploration of signification, though I will not conclude my treatment of it in the present chapter. The subject will extend through later chapters, for as I noted above, signification is intertwined with the microprocesses of modeling in strategic art: signification occurs and is recorded as a direct result of the act of structuring and changing models (or artificial schemata) to reflect analysts' assignments of meaning and, as possible, the conditions of those assignments. Since I will not discuss the details of modeling before the treatment of the art of strategic analysis in Chapter 1, the particulars and strategies of signification, and an estimate of the depth to which I expect it can be explored in the near future, must await consideration until that discussion.

But now that we have a more detailed concept of strategic analysis, it is appropriate to continue the discussion of the measurement problem in several dimensions, including various concerns about signification, relationality, *post mortems*, and the measurement of real and estimated effectiveness. Specifically, there are several crucial factors which I will consider in the next few pages:

- **Problems of Real and Estimated Effectiveness:** there are certain basic difficulties in strategic analysis which are fundamental to the problem of effectiveness and its measurement. I have indicated some of these above, and they must be explored further.

- **Current Research into Problems of Human Inferential Processes and Strategies:** the ways in which humans structure and store their knowledge, recall it and carry out judgmental heuristics are obviously of major concern to the consideration of measurement problems (and also, of course, to the consideration of the art of strategic analysis). Some recent research is highly germane and must be reviewed. Certain findings not only deepen our understanding of the problems of real and estimated effectiveness but also illuminate our search for better analytic routines, those systems of procedures, techniques and modeling approaches (such as described in Chapter 2) designed to improve the chances for more effective analysis. Indeed, as Richards J. Heuer, Jr., has remarked, there is a distinction to be made between the imperative to design ways now to improve analysis by deliberately imposing new procedure and methodology to mitigate *already known* causes of ineffectiveness; and the deeper, longer quest to understand in more detail the processes of cognition involved in analysis. If we wait for the latter, we may wait a long time indeed! We already know enough — controlled experimentation has established sufficient insights — to permit significant improvements in current analytic routines and techniques, and we must give priority to such efforts. This perspective, which I share, results in a practical, immediate emphasis on estimated effectiveness and on relationality. It does not, of course, diminish concern with signification, *post mortems* and real effectiveness. It is more a matter of what we consider now to be in the foreground and what we consider to be in the background. Such judgments do not necessarily imply that one is more important than the other. In my judgment, in fact, too much emphasis on estimated effectiveness and relationality will constrain us from dealing with certain critical problems involving the *esprit* and motivation of analysts, factors which probably are of the greatest concern in the long run, since they have much to do with the prospects for developing more successful strategic analysis and a tradition of effective analysts. But of these latter concerns there will be more said in subsequent chapters.

- **Skepticism About Current Understanding.** Having reviewed current insights into human cognition and inferencing, and exploited them for help in designing cognitive technology and in adding to the sophistication of analytic art, we must nevertheless invoke a protective and proper skepticism about the durability of such insights, recognizing the probability of deeper understanding in the future.
- **The Primacy of Imagination in Strategic Analysis.** I will underscore my view of the overriding importance of the strategic imagination and the fact that its enhancement is the proper essential focus of performance measures.
- **Some General Limitations on Post Mortems.** Although the subject will not be closed to further discussion, I will specify some important cognitive processes and phenomena not likely to be captured in *post mortems* for some time. I will also touch on certain basic controversies concerning the problem of interpreting records of meaning, a concern obviously crucial to *post mortems*.
- **A Preliminary Idea of a System of Measures and Variables.** Finally, in the context of the above discussions, I will structure a framework in which the measures (effectiveness, readiness and maintenance) are linked to the variables (backlog, relationality and signification) to form what I consider to be a coherent measurement system which, though preliminary in nature, possesses considerable power. Arriving at this system is the end purpose of all the discussion in the present chapter.

Some Basic Problems of Measurement

I will begin by insisting that all problems in measuring performance in strategic analysis should be examined in terms of the relationship between real and estimated effectiveness. To review briefly: real effectiveness is determined *a posteriori*. The strategic analyst pursues the future; real effectiveness must be sought after the fact through *post mortems*. In the meantime, however, ongoing strategic analysis must be performed and managed on the basis of *some* form of estimated effectiveness. *Always estimated effectiveness can only be based on what we think is the best present model or scheme of successful analysis; that set of techniques and procedures which, in our estimation, is most likely to promote effectiveness.*

The fundamental difficulty is this: since we have so much to learn about the process, techniques and methods of strategic analysis, and therefore so much to learn about how to design memories which enable us to create illuminating *post mortems*, we are now, and for some portion of the future, facing great difficulties in discovering powerful, pervasive techniques leading to real effectiveness. Indeed, we have yet to *experience* the intensified strategic analysis that cognitive technology will make possible. We are wise to judge that we simply do not yet know enough about what we are trying to preserve and measure.

The measurement problem is complicated by the fact that obviously strategic analysis cannot be a science. Always there are naturally unprecedented, unique facets to new strategic analysis problems. Hence yesterday's successful analyses are not necessarily suitable bases for learning ways to deal with tomorrow's problems: what we learn is never a final basis for estimated effectiveness. In fact, to forget this condition is to invite surprise. Therefore any analytic heuristic, though hopefully it will continuously improve, will never fully guarantee realism since the reality it pursues is an endless procession of unprecedented transformations which tend to elude our recognition powers, arising as these powers do from our experience of the past. Hence at any moment this imperfect temporary heuristic is simultaneously the ultimate criterion of estimated effectiveness and a defective, transient vehicle of analysis in the endless pursuit of real effectiveness.

Problems of Real and Estimated Effectiveness. As I said in my discussion of signification, we must attempt to determine real effectiveness through *post mortems* in which we review records of past meaning and the rationale behind the meaning, asking how and why we performed analysis a given way; and how and why we were successful or not. There are several levels of inquiry. First, there is *analytic accuracy* itself. Here the quintessential question becomes: Did we accurately project the outcome of a situation? If right, were we right by accident? In what dimensions or particulars were we right? Did we "miss" some processes but accurately project the major outcome? Most importantly, did our analytic techniques and methods of analysis, and the character of the analysis as influenced by them, directly contribute to success? Hence the design and implementation of both the signification and relationality records become crucial.

With such inquiries we are simultaneously seeking to improve estimated effectiveness, since we are attempting to learn the efficacy of present techniques; the character and impact of psychophysiological and behavioral factors; and the nature of innovative analytic strategies. All these will allow us to gain precision in estimating effectiveness.

Here is a partial list of the problems we must attempt to deal with in *post mortems*, either directly or through development of hypotheses for testing to learn more about ways to achieve analytic effectiveness:

- **Mind Set.** Some of the manifestations of this problem were cited above. Its sources and their comparative influences must be successfully explored: bad procedure, cognitive limitations, inadequate information, organizational and institutional influences, etc.

- **Exhaustion.** We must diagnose cases in which cognitive exhaustion or other pressures on the analyst cause inadequate analysis. We must begin to settle questions of pace, intensity and rest. Indeed, some problems such as mind set may worsen the longer analysis is pursued at a given time. Hence one area of subtle strategy we may learn more about concerns the timing of analysis: understanding when analysis should be interrupted or pursued from another angle.

- **Inadequate Modeling and Hypothesis Development.** We must learn how to diagnose and prescribe for cases in which strategic analysts fail to develop sufficient alternative projections.

- **Unproductive Organizational Arrangements.** *Post mortems* must lead us to wisdom about how analysts might best be organized. Should they be organized in loose arrangements of relaxed professionalism with modest attention to rank and hierarchy? What about other structures?

- **Inadequate Analytic Routine.** In Chapter 2 I defined an analytic routine, one that must be refined through the practice of analysis. In Chapter 4 I describe a system of strategic analysis art comprised of modeling techniques and other procedures that complement the analytic routine developed in Chapter 2. We must anticipate the invention of increasingly sophisticated and effective analytic routines, ones that progress far beyond those presented in this book, as we learn more about the sources of effectiveness.

- **Improvement of Cognitive Fitness Programs.** Through *post mortems* we must also learn how to develop cognitive fitness programs for analysts. Can we strengthen cognitive capabilities? Can we devise programs similar to training programs in athletics and in the performing arts? Perhaps cognitive fitness will turn out to be a combination of stamina and inventiveness of imagination; mental toughness and the willingness to take extra, perhaps unpopular, steps toward challenging hypotheses; and experience and skill in performing various techniques of analysis.

Today important new research into the problems and prospects of human inferencing is occurring which we must consider if we are seriously to seek to measure analytic performance.

In a recent book of considerable importance — *Human Inference: Strategies and Shortcomings of Social Judgment* — Richard Nisbett and Lee Ross have clarified some crucial aspects of human cognition. In part the book builds on insights developed earlier by Amos Tversky and Daniel Kahneman in their pioneering work on judgmental heuristics, as well as the work of a number of other diagnosticians of man as an inferential being.

It seems to me that this work, together with the work now being done by Richards J. Heuer, Jr., both constitutes extremely important research and is highly pertinent to the problem of strategic analysis, especially its measurement. Heuer is examining the research of Tversky, Kahneman and others and relating it to classes of problems encountered by intelligence analysts. Key portions of Heuer's work have been cited above; an important body of additional work is in process, some recently published in articles and some to be published in a book-length study. In particular, the following discussion draws from both Heuer and Nisbett/Ross.

Certainly current understanding of human inference — its nature, problems and prospects — must be utilized by those of us designing cognitive technology for support of strategic analysts. In simplistic terms, we now see man as possessing a painfully evolved and rather wonderful inferential capability: his judgmental abilities are indeed marvelous. But we must become more sharply conscious of these abilities; we must learn better when and how to use them according to certain strategies which may improve our interpretation of reality and our judgment of meaning. For we now perceive that in problems beyond pure science, man the inferential creature tends to rely too heavily and often with unfortunate results on certain intuitive inferential strategies, albeit strategies which are frequently highly successful; and to place too little reliance on formal, logical and statistical methods. One current model of human inference suggests that we depend enormously on certain preexisting "knowledge structures," that is, the schemata discussed above. These structures can be propositional; they can also take the form of complex scripts and be centered on personae. In using knowledge structures, we are believed to employ certain heuristics, most notably two: the "representativeness" and the "availability" heuristics.

Man's interpretation, understanding and representation of reality are, of course, greatly influenced by his existing knowledge structures and beliefs. Moreover, man tends to make judgments about reality through the use of that memorialized information most retrievable — that is, available — to him out of his total information base.

The knowledge structures and heuristics provide for efficient, effective inferencing processes; but they also create bias and cause faulty judgment. For example, experiments have determined that in assessing covariation we are often overly influenced by prior theories and insufficiently influenced by actual data configurations. These problems impede effective causal analysis and can seriously degrade projection. Certainly people ordinarily seem to have a tendency to adhere to theories rather than to seek disconfirming evidence.

Specifically, the availability heuristic is believed primarily to govern our attempts to determine the relations among events and the likelihood of events. Unfortunately, the availability of data in one's memory can be biased variously, and this in turn tends to bias our estimates.

The representativeness heuristic specifically is believed fundamental to that inferential process by which, given our knowledge of a particular situation, we estimate the probability that another situation may come into being. Another example is the judgmental task of categorizing various objects based on their characteristics and resting such judgments on what we believe are similarities between the understood characteristics of given objects and the presumed essential characteristics of categories. The problem is that mere similarity is often an unreliable guide.

Similarly, schemata or knowledge structures are at once indispensible for dealing with the deluge of impressions of reality and yet potentially misleading since for various reasons they can be or become inadequate models of reality. As I stated above in another context, they can, for example, become atrophied. Yet some now believe that judgmental heuristics may determine which knowledge structures we recall in given situations and how we apply them.

Certainly another (and related) troublesome element in human inferencing is the "vividness" factor: we seem to assign importance to information in relation to its vividness, its emotional interest, concreteness and imaginability; and its "proximity" in sensory, spatial and temporal terms. Vividness is seen as crucial to availability: more vivid information will be more memorable and hence comparatively more available for recall, leading potentially to the biasing of inferences and judgments. Unfortunately various kinds of very important data may take forms which are not vivid, for example, statistical forms. As Nisbett/Ross point out, although the vividness factor may sometimes be vital to successful inferencing, "the policy of weighting information in proportion to its vividness is risky."

These constraints obviously must greatly affect our ability to perform causal analysis and prediction. Doubtless we must begin the process of becoming more self-conscious about, more aware of, our inferencing strategies. Clearly this need is reflected in the twin goals of effectively performing strategic analysis and measuring that performance. Certainly the same need must inform our designs both of analytic techniques (such as the procedures and methods systematized in the model of analysis in Chapter 2) and of *post mortem* memories. Signification and relationality are especially pertinent here.

I must emphasize that a wise perspective in both the work of Heuer and of Nisbett/Ross is an obvious respect for the human inferential system of heuristics and knowledge structures. I share that respect. It is foolish to criticize the structures and dynamics themselves; rather, we must become much more aware of the frequent human tendency to use the structures and heuristics in self-defeating ways. A few statements by Nisbett and Ross on specific problems of human inferencing will suggest the value of using the insights of modern cognitive research in designing the cognitive technology of strategic analysis:

On Causal Analysis

Causal analysis is influenced strongly by two versions of the representativeness heuristic. One of these is a primitive requirement that the features of any putative cause resemble the features of the effect to be explained. The other, more normatively appropriate version, is the requirement that a putative cause resemble a causal factor in a theory explaining effects of the type in question. People often rely on poorly justified causal theories of questionable origin and place too much confidence in even those explanations prompted by causal theories held with good justification.

Causal analysis is also overly influenced by the availability of various causal candidates. People's explanations for events are therefore at the mercy of arbitrary shifts in the perceptual, verbal, or memorial salience of potential explanatory factors. This tendency, in turn, rests on an overly simplistic or "hydraulic" view of causality. People often seem to believe that a given event can have only one sufficient cause.

On Prediction

People do not seem to utilize population base rates in many prediction tasks and instead greatly over-utilize the representativeness heuristic; that is, they match the features of the target with those of the outcome and predict that the target will have the outcome to the extent that the target resembles the outcome. Similarly, people make generally nonregressive predictions for continuous variables. They tend to predict that the target will be as extreme on the outcome dimension as it is on the predictor dimension.

On Stubborn Beliefs

Belief perseverance sometimes seems to occur because people have an emotional commitment to the belief. Perseverance is likely even when there is no such investment, however, because (a) people tend to seek out, recall, and interpret evidence in a manner that sustains beliefs, (b) they readily invent causal explanations of initial evidence in which they then place too much confidence, and (c) they act upon their beliefs in a way that makes them self-confirming.

I have said that both for the effective measurement of strategic analysis and for its improved performance, we must sponsor in the analyst greater self-awareness and greater consciousness of analytic processes. Of course, the question often arises: What real accessibility to the workings of our minds do we have by any process? Recent experiments by cognitive psychologists appear to raise serious questions about the "privileged epistemic" power of introspection to discover causes of our mental status. Further, the literature of inspiration — the accounts of some great discoveries in mathematics, for example — seems to support the view that perhaps such processes are and will remain not merely mysterious but inscrutable.

Yet there is the view that language is a means of reflecting or "shadowing" our mental processes. It is evidence. It is a revealing record. If we accept this idea and extend it to include a variety of models — that is, put forth the notion that an analyst's models of situations, their conditions and dynamics, constitute such evidence, comprise such a record — then one of the sources of confidence in the present approach is evident.

Moreover, there is an additional and crucial consideration: it seems to me that an unprecedented extrasomatic memory of *complex* analytic modeling at least offers significant new hope of further enlightenment about human mental processes. In short, it seems obvious to me that we must look forward confidently to possessing better and better evidence, clearer and clearer records, with which to learn about cognitive and inferential processes. The records I speak of in the present book will perhaps have different textures, intensities and scope than we have hitherto had available to learn from, and hence may enlighten us considerably.

It is also important to consider the extent to which judgmental and inferential errors in analysis are based on *motivation* as opposed to cognitive and data factors. Again, the research would appear to suggest that if we consider motivation in terms such as "ego survival," motive may not directly distort perception or judgment. However, *behavior*, which is linked to motive, may be the mediator; it may have the major impact by changing the nature of the information available for recall. Nisbett and Ross point out that attempts to prove that motives directly influence perception, cognition and inferencing have not been notably successful. One problem has been to show convincingly that causes other than motive are *not* at work; and to show convincingly that motivation influences any responses other than comparatively superficial "overt responses." As Nisbett and Ross point out:

> . . . the *persistence* of unwarranted stereotypes despite logical or evidential attack seems quite explicable by nonmotivational processes. These include the assimilation of available information to preconceptions, the formation of causal theories, and the bolstering effect of belief-relevant behavior, all of which conspire to give stereotypes the illusion of impressive empirical support.

This outlook is important with respect to the view taken in these pages: that cognitive technology — the machine, the extrasomatic memory and the rigorous procedures of disciplined strategic analysis — can foster better analysis; that what is important is the faith that such attacks on the fundamental problems of strategic analysis will lead to important improvements.

I have listed above a number of pitfalls in analysis, and I have outlined portions of a contemporary model of human inferential abilities and limitations. These contributions to our knowledge are of considerable value, and yet we should entertain at least some skepticism about current lists of cognitive problems and about current models of inference. Do they point us toward real causes of ineffective analysis? The question arises because we have not really experienced the intensified, expanded analysis that cognitive technology will bring. It is conceivable that previous experiments in cognitive psychology and in other fields which seek to promote discoveries about how we think and imagine may not ultimately be of great use in understanding and measuring strategic analysis. In short, we must begin to practice the intensified modeling that we foresee, and create indicative records of it with which to explore underlying processes of analysis, before we will comprehend more fully the problems of analysis.

But for now one conclusion seems to me both inescapable and profound: in general, the fundamental difficulty in strategic analysis probably lies in *failures of imagination*. Strategic analysis is not basically a scientific process, and the imaginative development of increasingly sophisticated scenarios and models becomes essential. The fundamental challenge is the *imagining of realistic processes and outcomes.* In short, the important concept is that of the *strategic imagination.*

If we think of the many important problems of informal prediction we face in our daily lives, for example, attempting to predict whether a decision that affects our job security will be favorable, we will recall that frequently we may believe we have identified the key variables, made the right assumptions and modeled the decision process; yet frequently we learn after the outcome, often surprisingly, that we failed to *imagine* the causes and outcomes that did occur. In part this surely results because we spend comparatively limited time on such analysis, do so informally and without great discipline, and have no technical support.

Even in the remote future, however, we cannot expect to model future situations in their fullness; but certainly we should learn how to make fewer mistakes in analysis. We will learn largely by performing *post mortems* to discover the sources of effectiveness or its lack. Indeed, pursuing strategic analysis in this fashion promises to provide important insights into human cognitive processes.

But ultimately, the basic thrust should be to develop the strategic imagination and to measure that progress.

I will have much more to say about the strategic imagination in Chapters 4 and 5.

In discussing the pursuit of real effectiveness in terms of analytic accuracy, we must reckon with those crucial forms of human cognition probably too elusive for capture in the early analytic memories we design. Such difficulties obviously create significant design challenges which may seriously affect the early measurement of signification and relationality, and hence of effectiveness, both real and estimated.

Examples of mysterious and elusive, but perhaps crucial, aspects of cognition now seemingly difficult to record for *post mortems*, and hence for the discovery of the sources of real effectiveness, include aspects such as: the role of "background" information and perspectives in human cognition and analysis; certain *apparent* processes of the imagination; the general phenomenon of human perception of reality denoted, "gestalt;" states of "fringe consciousness" and their influence on perception and analysis; and various so-called "global" forms of human information processing. (Interestingly, an especially useful discussion of such problems is to be found in the work of Hubert Dreyfus on the limitations of artificial intelligence.)

In seeking to discover the foundations of real effectiveness, we need essentially to develop a memory in the machine which, though surely it cannot duplicate them, at least reflects key workings of the human mind in strategic analysis. In accepting this challenge, however, we are confronting such mysteries as the obscure semantic processes which allow us much of the time to interpret events, texts and sentences with little ambiguity. We are trying to understand phenomena such as fringe consciousness, a crucial level of human awareness exemplified variously, for instance, in our ability as we search for an individual in a crowd to develop simultaneously a certain and growing awareness of the appearances of other faces in the crowd. Indeed, our flow of experience throughout life creates fringe consciousness, for example, some years ago a man may have played football on the floor of a certain stadium, and years later as he looks at the front of this stadium he also "perceives" at the fringes of his conscious perception, the interior of the stadium. Clearly we seem to have perceptual and analytic powers involving cognitive abilities to be acutely aware of the background to a given problem, as if we notice out of the corners of our eyes aspects or features which may be far from prominent but are unquestionably influential; or as if we recall in the backs of our minds aspects or features which may strongly influence how we perceive and analyze a given object sharply in focus.

Presently it seems wise to assume that cognitive modes such as fringe consciousness and backgrounding are important influences on the ways in which we become alert to, and focus on, features of analytic problems. They very probably affect our judgment of whether the features appear dangerous, insignificant, interesting, etc.

And there are other processes of interest. For example, the strategic analyst will employ what appear to be certain *global* forms of developing alternative models, hypotheses and situation outcomes. These processes may *not* involve formal, "linear" analysis; they may occur "on the fringes" in subtle ways, yet influence greatly how we arrange our experiences and past judgments in order to analyze a present problem. As Dreyfus points out, human chess masters indicate that they experience at least two kinds of analysis: "zeroing in," by which they refer to a mysterious process of focusing on a problem or area known earlier in fringe consciousness; and "counting out," an elusive process by which the chess master effectively considers explicit alternatives, obviously by means other than the machine approach of reviewing thousands of options.

Consider also our capacity for ambiguity tolerance. As I noted, the semantic processes by which people interpret information are rather mysterious. When we use symbols we may be unaware of why, or under what processes, we grasp the intended meaning of the information. Context becomes extremely important: many instructions and other messages have a fuzzy or relatively ambiguous meaning, depending to some extent on the context or situations in which they are used; hence meaning must be made relatively unambiguous precisely through recourse to context and situation. But we know that humans have the ability to develop a "global context" which allows the establishment of meaning without having to eliminate ambiguity altogether. Given our capacity for ambiguity tolerance, it will be enormously challenging to develop memories which capture enough of the context to allow adequate understanding of the analytic processes involved in signification.

Finally, in the context of the problem of *post mortems* it is interesting to consider briefly that there are competing theories of the interpretation of records (a field designated, "hermeneutics"). In fact, there remains an arcane, highly technical and often confusing argument among theoreticians over interpretive limits. Some schools such as the Nietzchean and, later, the Structuralist, tend to deny that in certain senses we can recover original meanings of past records. Certainly this problem bears on *post mortems* and the determination of real effectiveness. Given that a record of analytic meaning would be created and stored in the machine, we may consider interpretation of that record from the traditional standpoint of textual interpretation. The memory of analysis would be based on a system of signification that would produce a "text" of historical meaning from which we would attempt to recover the original intended analytic meanings and their sources and associated rationales. Problems in such interpretation derive in part from the passage of time which causes new observers to review the previous analyses through their own new perspectives, thereby (it is maintained by some) obscuring the original meaning. Taken to its extreme, this view results in the position that all such interpretation is endless, futile, but perhaps important (and certainly fun) if you concede that new generations of readers should use significant literature of the past at least as a vehicle for viewing the reality of their own time. While few would not concede that some meaning in past records, especially very old ones, is probably irretrievably lost, a number of formidable observers stress the commonsense approach that there is obviously a great deal of objective meaning readily recoverable, particularly if the system of signification fosters the recovery.

In addition to accuracy, another level of real effectiveness involves *the impact of strategic analysis on decision makers and on decisions.* Can we determine whether and to what extent a given strategic analysis causes a decision maker to define and implement a policy? Obviously the relationship between the strategic analyst and the decision maker is crucial. It is, however, beyond the present scope to develop at any length that relationship; the connection between modern strategic analysts using computer-based aids and the decision maker and his staff, who will also use computer-based systems, is properly the subject of a separate study. Keeping in mind the final objective of developing measures of real effectiveness that involve the relationship between the decision maker and the strategic analyst, I will limit myself here to considering accuracy alone as the primary criterion for effectiveness. I am well aware that failure to account ultimately for the requirements of the decision maker is to risk unduly emphasizing analytic accuracy at the expense of analytic utility: they are not strictly the same thing. But we can see even now that basically the same approach as that taken to determine effectiveness in terms of accuracy applies to measuring the impact of analysis on decision making. It will be necessary to record and preserve in machine memory elements of the decision making process through which analytic perspectives are absorbed and used. Here again we must review cognitive processes, many very subtle. Hence there are notable analogies between the two cases which must be considered in the future.

The next and most difficult level at which real effectiveness must be measured is *the impact of strategic analysis on actual situation outcomes.* There is little to say about this challenge. Here we would be attempting to determine the impact of a piece of strategic analysis on an outcome such as de-escalation in a crisis situation. At this level obviously we confront very large and subtle problems. The ability to estimate the impact of a piece of strategic analysis on decisions and in turn judge the impact of those decisions on outcomes is to confront fundamental and well-known problems of history and evidence. For example, it may be impossible to obtain sufficient insight into operations in foreign governments to determine with assurance the influence of certain U.S. strategic analyses and ensuing policies. Or such insights may come only after considerable time has passed and historical records become available. Obviously one of the great challenges in *post mortems* is to determine effectiveness at this level as completely and rapidly as possible. Equally obviously, one can only state that often good fortune is required for even the opportunity to make such determinations. Yet they are far from impossible, and an effort must be made to achieve them whenever they are feasible.

A Viewpoint on Estimated Effectiveness. From the above discussion, several points seem especially to bear on estimated effectiveness: First, real effectiveness is at bottom concerned with the effectiveness of analytic technique, of analytic methods and procedures.

Second, we are now in the infancy of such technique. Until the advent of information science and cognitive technology, we had not really begun to discover a tool, a weapon, for grappling with the problem of strategic knowledge and analysis. We should assume that our understanding of technique will grow.

Third, the first two points seem crucial to considerations of estimated effectiveness. Estimated effectiveness is a wager that existing analytic techniques and procedures will produce rigor which in turn leads to real effectiveness; but as we know, current technique is primitive. Moreover, the sophisticated determination of real effectiveness must await the *post mortem*, and it must be assumed that over the near term *post mortems*, like technique, will be comparatively primitive.

Fourth, we must assume that as we learn how better to do strategic analysis through the more intensified analysis and secure memory afforded by the machine, master analysts will appear, innovators of technique who will carry us much further toward analysis that promises effectiveness. In a very real sense, we simply must patiently wait for these analysts. I will speculate about them in Chapter 5 and sketch imaginary portraits of two of them.

Fifth, we have already pointed out that each new strategic analysis problem is unique; that by its nature each has a novel character which challenges strategic analysts. Therefore, we cannot ever anticipate entirely trustworthy measures of estimated effectiveness based on the "lessons of the past."

All these factors point to the issue of the proper *present* perspective on estimated effectiveness. A year ago I might have judged that at least in the near term, we know too little about the strategies and tactics of analysis to believe we might profoundly understand such conceivable measures of estimated effectiveness as *elegance* and *economy* of modeling activities; the concept of *actionability* — that the product of analysis be highly usable for the decision maker; and *productivity* of analysts. Yet my colleague, John W. Sutherland, the acclaimed systems scientist, has recently begun to attack the problem of estimated effectiveness and has produced early results which seem to me very promising. The work is part of some funded research and development Sutherland, the present author and others are currently performing for the Defense Advanced Research Projects Agency. Although these and later results must be reported in detail elsewhere when the projects in question are completed, I will note some very general and preliminary features in the emerging design:

- **Multiple dimensions of estimated effectiveness.** Several aspects of estimated effectiveness have been identified and studied.

- **Classification of analytic activities.** Technical definitions of various types of analytic activity have been developed which encompass and refine the relationality and signification variables.
- **Definition of levels of strategic analysis.** The anatomy of strategic analysis has been explored further and several levels of analysis identified. These in turn have been related to various types of analytic models (artificial schemata), some of which are described in Chapter 4.
- **Strategic situation categories.** A hierarchy of threat situations analysts confront has been structured. This process has led to the definition of several generic categories of strategic situations which vary considerably in analytic difficulty. These categories, together with the levels of strategic analysis, become very important in the derivation of estimated effectiveness and in the diagnoses essential to *a priori* evaluation of analysis.
- **Productivity functions.** Functions have been defined for the productivity of strategic analysts. Among other things, these functions are extremely important with respect to the evaluation of the analytic process and the potential for improvements through the application of new technology. The productivity functions are also obviously crucial to estimated effectiveness.
- **Baseline productivity models.** Models of productivity have been developed which take into account the size and complexity of given types of analysis with respect to different levels of analysis and different degrees of difficulty. Again these models are very important for the management of resources in a strategic analysis center and for the development of baseline data for evaluating different portions of the process.
- **Computational effectiveness models.** Work progresses to develop models for accuracy, readiness and other elements of effectiveness.

I mention these developments to underscore the fact that despite the great distance we have yet to go in achieving a powerful understanding of estimated effectiveness, we have already begun to make significant progress toward that understanding. The importance of this progress lies in the fact that estimated effectiveness becomes the basis of day-to-day management of analysis. How else does one determine whether one is managing limited resources in strategic analysis in a judicious fashion? Thus we must *immediately* seek to develop reliable measures of estimated effectiveness. In short, even if our expectations for the near term are modest, we must take the quest for these measures very seriously.

Against this background, I will now explore the relationships among the measures and variables.

The Measures and Variables as a System

We have considered the measures and variables individually but their full definition resides in their linkage, interrelatedness and interdependence. In the context of the constraints discussed above, we must now explore how the basic analysis measures — maintenance, readiness, real effectiveness and estimated effectiveness — connect to the key variables of backlog, relationality and signification; and how this creates linkage among the measures.

Table 4 is a matrix summarizing these interrelationships. Obviously the matrix must be considered a preliminary framework of measures — something of a metaphorical scheme — especially in view of the problems I have just discussed.

Commentary on Table 4. As shown in the table, *maintenance* is associated with four variables: preventive maintenance (PM), backlog (BL), relationality (R_T) and signification (S_T).

The maintenance measure depends, first, on monitoring the time that support systems are used versus the vendor repair and system downtime schedules, which provides preventive maintenance measures. Second, it requires measuring backlog, derived by determining for given times what data in various areas has not been analyzed within the various analytic stages.

Third, maintenance entails measuring relationality. In relationality one factor taken into account is the time spent by analysts in relating given data to given models, in following procedures and in employing other analytic techniques. For a number of reasons managers and analysts might wish to monitor and compare such values. One reason can be suggested by drawing an analogy to the analysis of intelligence imagery. In some analysis of aerial photography, the objective may *not* be to do detailed, painstaking review of various photographed scenes, especially if the photography does not have adequate resolution for extremely fine analysis. The purpose of analyzing such low resolution imagery might be to enable search of areas to determine possible targets which subsequently should be monitored in higher resolution imagery. The analytic time spent on a single frame of high quality imagery of a significant target ordinarily would be greater than on a lower quality image of the same area. Similarly, different stages in the progressively more difficult challenges over the sequence of strategic analysis from monitoring to projection should result in variations in times reflected in relationality. With enough data, basic trends and characteristic time patterns might begin to emerge as a set of analytic norms. Such data might be very

Table 4. Interrelationships Among Measures and Variables.

Analysis Measure	Variable	Data To Be Monitored	Calculation
Maintenance (M) (an activity)	Preventive maintenance (PM)	Time used versus vendor schedule	$PM = \Sigma\ ts$
	Backlog (BL)	Data on given geographic areas (or other problem categories) not analyzed at given stages in analytic process	$BL = \Sigma$ data awaiting analysis in the various analytic stages
	Relationality (R_T)	Analytic models/procedural routines involved versus total available data Time spent on each model (T_i)	# models $R_T = \Sigma\ T_i$
	Signification (S_T)	Changes to variables at analytic nodes Account taken of analyst skill (A_j)	$S_T = \Sigma\ S_i \cdot A_j$
Readiness (R) (a state)	NA	NA	Scaling based on PM, BL, R_T and S_T measures (ideal M versus real M) Objective/subjective reference points on scales
Estimated Effectiveness (EE) (a result estimated *a priori*)	Forecast probabilities as determined by analyst	Forecasts in projection stage of analysis	$EE = \Sigma\ R$ + forecast probabilities

Table 4. (Cont'd)

| Real Effectiveness (RE) (a result sought to be determined in *post mortem*) | Forecast probabilities as determined by analyst | Forecasts in projection stage of analysis. Subsequent real events in projected period [via *post mortem* (PT)] | $RE = \Sigma\ EE + f\ (PT)$ |

important for diagnostics: for example, in operations analysis and research and development activities designed to afford analysts more time to do the more difficult parts of analysis and to reduce time delays in parts of the process which should be done more quickly. Another purpose in examining comparative times is to build in a safeguard against attempts to reduce backlogs by skimping on conducting the relational aspect of analysis. Moreover, the time factor ultimately could more sharply delineate for managers dilemmas created by limited analytic resources. But obviously caution and empirical wisdom are needed since there are not necessarily ideal times for performing various portions of analysis. Clearly, analysts will differ in ability, problems will differ in complexity, and data will differ in their applicability. In the final analysis, the ability to monitor times spent in relationality and in reducing backlog might aid in the general effort to ensure that a variety of questions of a diagnostic nature, many not now fully anticipated, could be answered in the future.

Signification values are also determined for the maintenance measure. Note that the calculation of signification includes a value, A_j, which refers to the estimated skill of the analyst(s). This is an experimental factor. "Skill" could be derived from experience, training, demonstrated interpretive success, etc. Empirical data and a period of testing the practicality and utility of this factor are required. It would prove interesting to attempt to determine correlations between different analyst profiles and instances and degrees of signification. In the case of specialists who largely (but not exclusively) operate the system, the A_j factor would be of marked interest.

Readiness, an analytic state, is expressed in the form of scaling based on PM, BL, R_T and S_T. Readiness is basically derived by comparing ideal maintenance to real maintenance. Note also that under the calculation of readiness there is mention of objective and subjective reference points on the readiness scales. Strategic analysis managers must allocate limited analysis resources to a wide range of monitoring responsibilities. Managers may assign different goals of readiness to different problem areas and different cadres of analysts.

Under *estimated effectiveness*, it is seen that the analyst will determine projection probabilities (this is illustrated in considerable detail in Chapter 4) and that for projections readiness will be factored in as a means of determining estimated effectiveness. In sum, current readiness should qualify projection probabilities and produce estimated effectiveness.

Real effectiveness, as discussed above, is ultimately determined through *post mortems*. In Chapter 4 the problem of measuring real effectiveness is considered in more detail, and some strategies are developed.

It is necessary to consider analytic art -- procedures, forms, and methods -- in more substance before addressing further the issue of real effectiveness.

The diagram shown below indicates the present level of conceptualization of the full system of analysis performance measures and their relationship to the key variables.

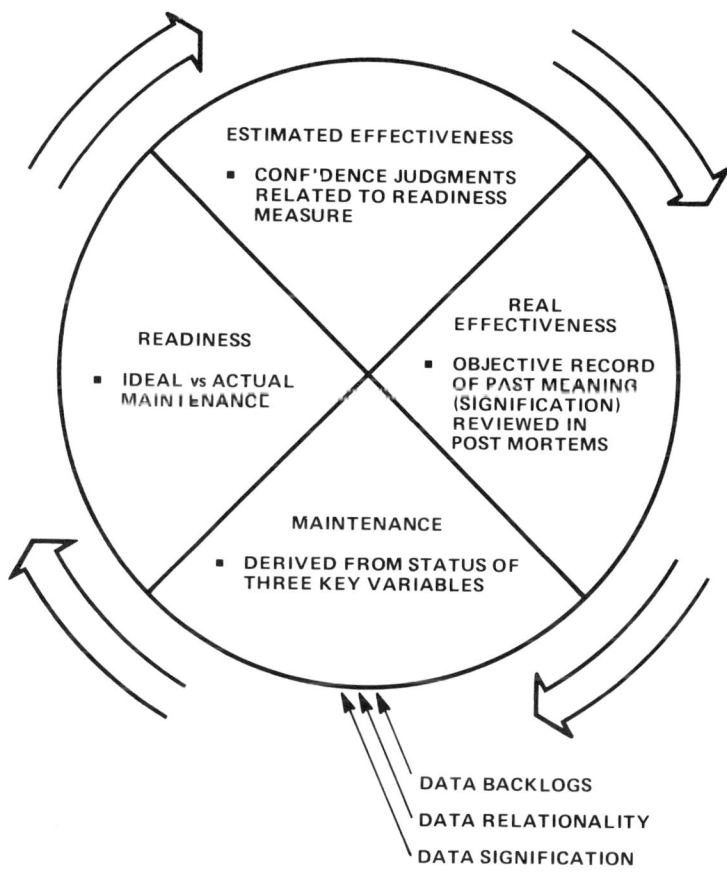

RELATIONALITY: MEASURE OF EXTENT TO WHICH INPUT DATA IS ANALYZED BY RELATING IT TO SYSTEM OF ANALYTIC PROCEDURES, MODELS, METHODOLOGIES

SIGNIFICATION: MEANING ASSIGNED TO DATA BY ANALYSTS. CAPTURED BY RECORDING CHANGES AT CERTAIN NODES IN ANALYTIC PROCESS (e.g., CHANGES IN VARIABLES IN THREAT MODELS) AND ASSOCIATED RATIONALE

76 TOWARD EFFECTIVE STRATEGIC ANALYSIS

The definitions which now appear in the wedges within the diagram under the four basic measures are explicitly related to the key variables. In addition, the four arrows positioned at different points along the circumference of the circle are meant to indicate the linkages among and across the measures and variables.

Table 5 makes more explicit the linkages between maintenance and readiness, and between readiness and estimated effectiveness.

Table 5. Measures Linkage

From	To	Calculated By
Maintenance	Readiness	Measured readiness times a function* of maintenance activity/effort required to bring maintenance up to 100 percent for specific problem area or stage of analysis. $R \cdot f(M100\% - Mcurrent)$
Readiness	Estimated Effectiveness	Measured estimated effectiveness times a function* of the differential between measured readiness and optimum readiness.

*Functions must be refined on the basis of empirical data.

Note that the link from maintenance to readiness is based on the concept that a certain degree of effort and activity must be identified that would be required to bring maintenance up to 100 percent for any analytic area or stage of analysis at any given time. (Empirical wisdom is again essential here.) This perspective would be important for the manager who must allocate limited resources in responding to problems of varying priority.

Subvariables. Many perspectives can be gained by collecting data on the three major variables. For example, a number of subvariables relate to the major variables. Examples include:

- **Types of data.** One can consider different types of information and how they are being analyzed at the different analysis stages.

STRATEGIC ANALYSIS MEASURES 77

For example, we can consider content categories such as political, economic, military, etc. In short, information may be categorized by various criteria and monitored thusly.

- **Percentages.** It will be important to determine percentages of information at different levels of the analytic process. In turn, percentages can be related to types of information or can be related to other parameters.
- **Capacity.** Such values as data capacities at various levels of the system memory, or data capacities with respect to throughput capabilities at various nodes, might be of interest.
- **Time.** One can monitor time variously: variables include durations of certain processes; speeds of certain functions; rates of certain activities, for example, rates of backlog reduction; and accelerations.
- **Forms.** The forms that certain types of data tend to take and then lose during different processes within the analytic system could be of interest.
- **Gaps.** Data gaps and activity gaps in the process of analysis could be of considerable interest in terms of various measurement problems.

States and pathologies. We can think of a strategic analysis system as exhibiting different states. Measurement data may eventually allow us to explore acceptable steady states. To define such states, we would need to determine heuristically relationships between input, throughput and output capabilities and factors such as the number of analysts, the state of the technology support, etc. Such studies might eventually provide considerable insight into technological solutions for problems in readiness and effectiveness.

Pathological states in information systems occur when one or more variables remain for a significant period beyond their ranges of stability; or when the costs of adjustment processes required to keep them within their ranges of stability are significantly increased; or when the changes from an efficient state are caused by either malfunctioning of subsystems and components or unfavorable conditions in the environment.

Major categories of causes of pathological states in information systems include the following:

- **Excesses of information input.** The problem of excessive information is, of course, highly germane to strategic analysis. Adjustment procedures on the part of analysts would be of key interest to management. Excesses in information input could cause delays, loss of focus and create inappropriate emphasis on certain portions of the analytic process to the relative neglect of others. All this

in turn relates closely to the epistemological and cognitive problems which beset analysts in any event.
- **Lack of information input.** Lack of information input is also very relevant to strategic analysis. We have discussed this problem in terms of epistemological and cognitive obstacles to effective analysis; one can see the relationship to the analytic process as a system. Indeed, the lack of information could increase the risk of bad analytic judgments.
- **Inputs of erroneous or mischievous information.** The effects on the strategic analytic system of the input of deceptive data, erroneous data and data of relatively poor credibility obviously could become serious obstacles to effectiveness.
- **Abnormalities in internal information processes.** Such abnormalities must be identified and corrected in a system of strategic analysis. Specific pathologies could involve the transmission of information, the coding of information within the system, the associating of information, decisions concerning the meaning of information, and other activities of information processing subsystems. Examples could include semantic problems; errors in routing information; and information that is poorly stored in memory, forgotten, not readily retrieved, and hence not *available*.

In relation to such pathologies, there is a set of classic hypotheses about the likely behavior of information-related systems under varying conditions; for example, when changes are introduced in volume, type of information, stress, etc. I am confident that the measures discussed above may facilitate useful research into some of these hypotheses.

Some Comments on the Uses of Measurement Data

The previous sections discussed the nature of the measures developed for strategic analysis. We should consider also the limitations on the measures data and explore appropriate uses of the data. Of course, definitive answers to both questions — utility and constraints — must await the collection and analysis of empirical data.

It does seem appropriate to assume, however, that managers would utilize backlog and relationality as aids in day-to-day managing of resources and evaluation of operations within strategic analysis centers and groups. With respect to signification, my perspective is somewhat different. Analytic difficulties caused by problems in backlog and relationality, which in turn are caused by factors such as resource shortages, constitute one type of difficulty, one where management responsibilities for remediation are clear. However, the roots and causes of an analytic failure might be shown to derive primarily from problems in epistemological and cognitive aspects of analysis, reflected in a system of analysis whose procedures, models and techniques were inadequate to the situation. Among other things, the signification measure is designed to illuminate these kinds of problems. Solutions to such problems will in part require conceptual and technical innovations and improvements rather than the sort of adjustments typical of everyday management. Although sharp distinctions among the sources of strategic analysis problems are not always likely, it would be useful to determine the relative importance of various sources of difficulty. Different problems will have different solutions, some of which will be possible given simply a decision (and the requisite resources) to make a change; others will require further technical insight. The real purpose of the signification measure, then, is to provide a basis for more productive inquiry into classes of subtle problems in strategic analysis. (Presumably it has been clear that the intent behind the signification measure is not the second-guessing of analysts.) Knowing about the problems, we must, of course, proceed to the formulation of solutions.

Transition

I have discussed pathologies in information systems and the importance of measures to diagnose them. Without measures we confront great difficulties in the face of symptoms of inadequate strategic analysis in the individual and in the organization. Certainly we know enough to expect that drastic symptoms will sometimes arise. Perhaps the most serious symptom, one whose pursuit is implicit in the entire measurement scheme I have devised, is indirectly suggested in these two related but noncontiguous passages by the writer, Joan Didion, commenting in *The White Album* on an illness she has suffered. I have juxtaposed the passages. Referring to certain organic disorders of the central nervous system, Didion writes:

> What happens appears to be this: . . . the lining of a nerve becomes inflamed and hardens into scar tissue, thereby blocking the passage of neural impulses . . . . During the years when I found it necessary to revise the circuitry of my mind I discovered that I was no longer interested in whether the woman on the ledge outside the window on the sixteenth floor jumped or did not jump, or in why. I was interested only in

the picture of her in my mind: her hair incandescent in the floodlights, her bare toes curled inward on the stone ledge.

In this light all narrative was sentimental. In this light all connections were equally meaningful, and equally senseless. I was meant to know the plot, but all I knew was what I saw: flash pictures in variable sequence, images with no 'meaning' beyond their temporary arrangement, not a movie but a cutting-room experience.

I close the present chapter with these beautiful passages because they describe powerfully a pervasive experience, one that seems almost native to our time, which is far removed from the successful pursuit of strategic art, a subject to which we now turn.

4
The Art of Strategic Analysis

We must now focus on the art of strategic analysis. The process of analytic art is what we seek to illuminate, measure, manage and improve. In attempting to meet these goals, I have designed a computer-based system of strategic analysis which both enhances the pursuit of analysis and facilitates the implementation of performance measures. By way of introduction, let us examine quickly two aspects which we will subsequently explore further: first, some major problems in strategic art; and second, some major features of cognitive technology which are intended to mitigate the problems.

Major Problems. Important constraints on strategic analysis include the following:

1. To begin with we can see several *fundamental problems* confronting the interpretive strategic analyst in pursuit of strategic art. In fact, these problems have created a crisis of interpretation. The first problem was discussed above: *the natural limitations of human intuitive inferential abilities*, the frequent traps in judgmental heuristics and in the nature and use of knowledge structures. Two other major problems discussed in some detail below are: certain *limitations of science and formal methods*; and *the ironic demoralizing impact on analysts (and others) of the massive information deluge created by modern communications media*.

2. Certainly in a time-critical sense, the unaided human mind seems unable to process, the then conduct extensive interpretive cognition on, the massive modern information input. With reference to our model of strategic analysis, the analytic process is truncated. The basic result is a failure by analysts to mount a strong, extended interpretive effort.

3. For reasons discussed above, in the cognitive process of analysis we forget as we learn. Therefore we have great difficulty in establishing continuity and accuracy in the memory, the reconstruction through *post*

mortems, of meanings and rationales. This condition is continuously exacerbated by the intensity and peculiar features of the modern information deluge.

4. The mind must have recourse to models — schemata or knowledge structures — through which it filters the input data in order to make sense of it. But the problem described in item 3 has resulted in a failure to build lasting sophisticated analytic contexts for analysis and interpretation of new data. In short, it is a problem of the disappearance, the fading out, of analytic context.

5. The schemata arising in the mind are both inborn and learned. Learned schemata are the major analytic tool, a major response of the brain, in operating on indirect and secondhand data, which are the classic inputs to the strategic analyst who typically receives reports, messages and visual data on remote places, persons and events.

6. Four additional problems now ensue. The first is a problem in time. In the face of the persistent information deluge, there may be insufficient time — literally no pause — for the unaided individual human analyst to develop and, more importantly, *refine* appropriate schemata. Schemata are required that are refined enough at least to give promise of dealing with some effectiveness with the awesome challenge of strategic analysis: the interpretive task of attempting to understand remote parts of the world with literal realism, recognizing threats (including novel threats), and at least warning appropriately on probability. This is not to broach the more fundamental problem of the constraints on the unaided memory of reconstructing variations in past schemata.

7. The second problem is this: even if there were time, the *unaided* human mind, no matter how heroic and endowed with genius, probably would not invent numerous powerful schemata, for all the reasons given previously, all the constraints on schemata preservation and storage, and hence refinement, which arise from the nature of the mind's operation on data.

8. As a result of these two problems, the unaided mind has a third problem: difficulty in achieving a certain kind of contemplative stasis. As I have said, hindsight analysis and *post mortems* are key to analysis and its measurement and management. Management of analysis begins with the individual analyst managing his own analysis; obviously *post mortems* and hindsight analysis must not be the responsibility only of committees exploring analytic failures but should be part of analysts' day-to-day attack on analysis problems. For this there must be stasis, a freezing of past situations of signification, those occasions when the analyst (or analysts) assigned meaning by operating cognitively on the

data through specific rationales. But as we have seen, signification capturing will not be adequately performed by the unaided mind.

9. The fourth problem, which I shall discuss below, is the difficulty of the unaided mind in rapidly perceiving far-reaching ramifications (or implications) of strategic judgments. We might easily recognize that such judgments change single elements or a few elements in complex strategic problems whose modeling involves many elements interrelated in a coherent framework. But how readily can we then recognize that the same judgments also ramify to cause shifts in likelihoods at *many* other locations in the model structure? In the case of a complex set of interrelated schemata, the shifts could create structural changes at numerous points, with profound implications in meaning.

Such problems are sobering indeed. That their mitigation will entail a very long struggle is painfully obvious. There is surely no need to mediate on this fact in the present pages. Rather, I shall simply proceed to discuss what I believe are the beginnings of a resolution of the problems: future systems of computer-based strategic analysis art and their associated performance measures.

Some Features of Proposed Cognitive Technology. The following are critical features:

1. I have designed for the analyst a structure for a machine-based, working memory of strategic analysis. By memory I mean essentially a computer-stored, secure, selective context comprised of chosen portions of the analyst's accumulated analysis against which he analyzes new data to determine meaning. The memory structure consists of ten classes of *forms*. The forms constitute a set of artificial schemata. Highly changeable, they may be thought of not only as a structure for memory but also as a language for various kinds of modeling. The forms are comprised of certain types of flow diagrams, various kinds of mapping, and other diagrammatic structures. As such, they are especially well-suited for computer-driven displays. A machine-operated configurational language is provided by which the analyst can change, recreate, develop and display forms rapidly. Hence there is flexibility within the internal structure of the forms: idiosyncratic, subjective aspects of analysis are accommodated. Further, new meaning and rationale may be registered easily, and innovations in technique may be incorporated readily. Within the language of the forms, then, the individual analyst can perform, record and display his particular analysis. In short, he can extensively pursue the strategic art.

2. The ten forms each pertain to a different point in the model of the strategic analysis process. For example, four of the forms are used in the threat recognition stage; several others are used in the projection stage.

Further, the forms are sequential: one leads logically to the next, and so on, through the full process of strategic analysis as I have modeled it in Chapter 2. The forms are, in short, an interrelated, progressive series, a chain of outputs of different analysis steps and procedures. Therefore, the language of the forms has a syntax.

3. Together with the forms, I have designed an *analytic routine*, an itinerary of interrelated analysis steps whose results are localized, focused, recorded and displayed in the context of the forms. The routine is designed to insure that the processes of analytic art described in Chapter 2 are followed appropriately. Thus there is an overall grammar of analysis.

4. Hence the forms and the routine comprise a *system* of analytic art. It has been designed to enhance analytic effectiveness. It preserves, tailors and allows manipulation of a large operational context of prior analysis as a framework for assigning meaning to new information. It incorporates the use of antibias techniques and other effectiveness-sponsoring procedures. The extrasomatic memory based on the forms, together with the analytic routine, is specifically designed to work against certain cognitive and epistemological problems in analysis. Most important, the approach is designed to foster the vital heuristic tendency: *the system gives the analyst the means to change the system.* Hence the evolution of analytic art is facilitated by the system itself.

5. The performance measures are implemented simply by using the computer-based system. When forms are changed in certain ways, signification is evidenced and recorded. Signification is displayed and reviewed by recourse to the memory of forms. The memory of the forms becomes the data base for *post mortems*. By following the routine, and evidencing this by computer operations such as calling up forms, relationality is pursued, automatically monitored and recorded for derivation of measurement data. By a system of identifying and tracking the flow of specific input data through the routine and in relation to the forms, backlog values are generated.

The remainder of the present chapter includes a detailed discussion of the present system of analytic art. The discussion is organized around the stages of analysis: monitoring, threat recognition and projection. Each is discussed separately. For each stage, the following topics are stressed: the analysis forms; the analytic routine; the development of signification values; and aspects of cognitive technology.

Figure 8 shows some of the areas of emphasis in the discussion in the present chapter. Again, the less pertinent areas in the diagram have been partially obscured.

In undertaking these discussions, I wish to stress that the analytic

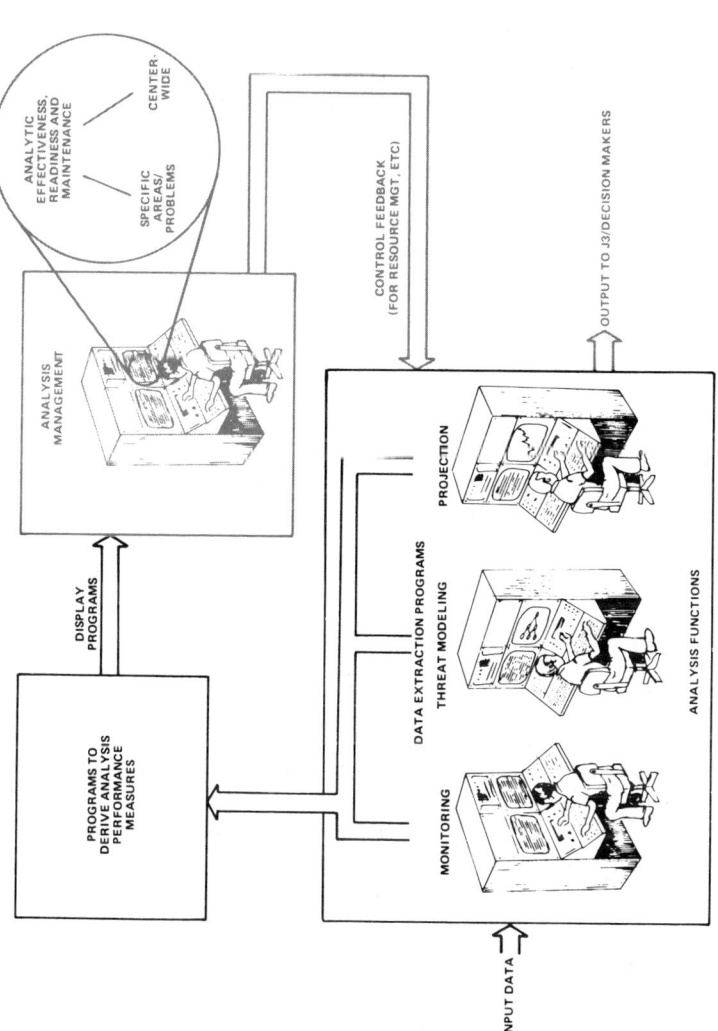

Figure 8. Methods Emphasis.

forms and routine presently and confidently considered are a first generation approach — indeed, a primitive one — that would be refined and changed by analysts. One of my major premises is the long-term progressive improvement of strategic analysis through the analytic heuristic facilitated by cognitive technology. I have not the slightest belief that the system of analytic art put forth below is either inclusive of all currently available techniques or other than a temporary approach.

Against this background, we may now discuss analytic art in terms of the individual stages of strategic analysis.

Monitoring Stage

In conducting the day-to-day watch on an area, the strategic analyst monitors the activity levels of numerous indicators. Typically there are preestablished lists of indicators of political, economic, military and cultural events. These are judged to be necessary and significant events in processes leading to the enactment of threats. For example, indicators may be developed and organized to show increases in the likelihood of an economic decline or in the readiness of military forces to launch an attack. Political indicators may be developed and monitored to provide early warning of potential international crises through detecting the emergence and character of political hostility between or among nations or factions.

Indicators are generally categorized. For example, military indicators may be subdivided into such categories as: unusual deployments of military forces; intensive logistics activity; military exercises conducted at atypical times; and similar examples.

There has been considerable indicator research directed to the warning and crisis management problems over the past decade. The research has led to definition and development of indicators and to design of computer-based aids which operate on large volumes of indicator data and produce quantitative output on activity levels, for example, the number of adversary ships in or out of position by specific geographic area. Given the input of indicator data from a variety of sources, the basic purpose of such aids would be to help the analyst determine to what extent, based on the norms inherent in accumulated data, there are unusual activities occurring; and the likelihood of various threatening trends developing. Judith Ayres Daly of the Defense Advanced Research Projects Agency describes one monitoring aid now under development which involves political indicators:

> The Early Warning and Monitoring System is a computer-based interactive global warning system. It is based on quantitative political indicators which, along with probabilistic forecasts,

are displayed in graphical or tabular form. (The system) is being improved by the addition of rigorous forecasting methodologies and automation of many of the necessary but time-consuming tasks which must be performed by (an I&W analyst). These include automatic generation of hotspot, alert, and monitoring lists which are designed to supplement the analyst's other sources of information and save him time by focusing his attention.

Clearly such approaches and technology tend to mitigate problems of information overload and to enable analysts to realize the opportunities inherent in the availability of large volumes of data.

Indicators should seldom be considered outside the context of situational models, for the analyst eventually must seek to make sense of the activity of indicators within some framework. If indicators are listed in a sequence believed to represent a probable chronology of necessary and significant adversary preparations for military attack, the indicators in effect have been assembled into a model. In the present research, threat modeling, which occurs in the threat recognition stage, should incorporate indicators associated with the monitoring stage. At this point, we should proceed to a consideration of the threat recognition stage and discuss it together with monitoring. The close relationship between the two will become clear.

Threat Recognition Stage

As the anthropologist, Edward T. Hall, has said, "Man is the model-making organism *par excellence*." In the threat recognition stage, the analyst now faces a much greater challenge to his imagination and general analytic ability, for he is attempting to model potential courses of action by various countries, entities and decision makers in his area of interest. These models act as filters through which he reviews input data, particularly data on the activity of indicators from the monitoring stage.

The most frequently used term for such models is *scenario*. The strategic analyst is indeed a scenarist. He must imagine hypothetical situations as an essential part of the process of analysis. He must use his experience captured in his memory, and the experience of others in a variety of forms, in creating the situations.

The analyst must be a model builder because the *situation* (not the event) is the most meaningful unit of strategic analysis. The term, *situation*, is of special importance. *Situations* are considered as "things" to be recognized by analysts. Situations may be defined as consisting of three stages: (1) the present; (2) processes leading to an outcome or situation result; and (3) the result itself. The analytic process should result in recognition of a specific threat situation in advance of its full impact. The process can be shown as follows:

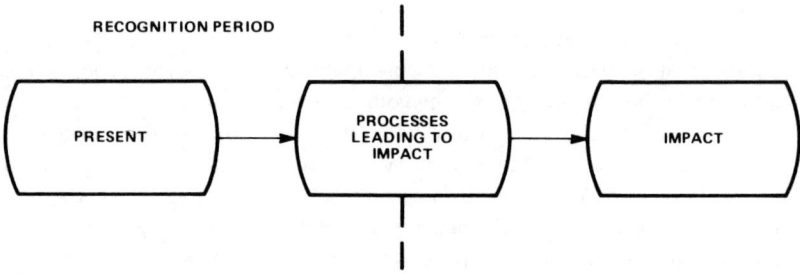

It follows from the primacy of the situation, itself a complex arrangement of components, that the analyst becomes a model builder, a scenarist: it is how he must envision and analyze situations. Here the concept of *artificial schemata* or *forms* becomes important.

Analysis Forms. We now turn to specific examples of analysis forms. A number of examples could be given, including examples centered on corporate problems, energy issues, economic conditions and national security and defense matters. It would be instructive to delve into each of these areas, but to do so in the present context is not possible. We must examine a single case, keeping in mind that fundamentally the problems of strategic analysis are in numerous respects highly similar across all such examples. As a hypothetical case, we will focus on a national security case, namely, the problem of warning of hostile activity directed against U.S. interests and assets by North Korea. A set of four interrelated analysis forms believed useful in monitoring and threat recognition has been developed as follows:

● **Projected Alternative Major National Courses of Action (PAMNACS).** This form is a structure for expressing what are believed the most definitive national policies and courses of action of a country such as North Korea with respect to key strategic aspects, such as reunification of the Korean peninsula and the international position of North

Korea *vis a vis* ROK, Japan, USA, Europe, PRC, USSR and other countries. Figure 9 shows an example of the PAMNACS form used for projection of some alternative major courses of action by North Korea. Several possible courses, ranging from negative to positive impact, are identified. Certain courses can be considered as leading to possible crisis situations. Note that the PAMNACS form is compatible with a system of computer-driven displays.

The PAMNACS should be as comprehensive as the analyst (or group of analysts) believes necessary. Thus PAMNACS are *always and readily changeable*. Figure 10 is a further example in which two alternative courses of action by North Korea are expanded beyond the first version of PAMNACS. The two alternatives are, first, that North Korea attempts to unify the peninsula by war prior to any U.S. military withdrawal (a withdrawal, say, of the scope contemplated early in the Carter administration); and, second, that the North Koreans seek to unify the peninsula by war after such a withdrawal. Let us imagine that a study has been made of the various forms such eventualities might take. These forms could include: a protracted conflict; a quick strike by the North against key targets in the South and the seizure of these targets, leading presumably to eventual demoralization of the ROK and its capitulation; and a limited strike by the North, for example, against offshore islands, designed to foment tension and unrest in the ROK which leads to a new government in the South willing to consolidate with P'yongyang. The specific examples given here are not as important as the principle that PAMNACS must be flexible; that it must accommodate new analytic perspectives as they arise.

- **Decision/Event Networks (DEN).** Figure 11 shows an example Decision/Event Network, the second type of form. DENs are extensions of the PAMNACS. For each of the possible major national courses of action shown in the PAMNACS, a DEN is developed and serves as a means of further modeling the possible activity that North Korea might exhibit in *implementing* that particular course of action. Figure 11 refers to a specific course of action by the North in which it seeks reunification with the South by military means after a U.S. withdrawal; and plans to do so by means of a protracted, all-out war. Figure 12 shows a DEN for the same course of action but under a different military plan: the quick strike option. Similar DENs would be developed for remaining elements in the PAMNACS. Of course, DENs are always changeable.

The events and activities in the DEN are developed on the basis of the indicators from the monitoring stage of analysis. One of the major purposes of the forms in the threat recognition stage is to provide a context for the indicators.

90

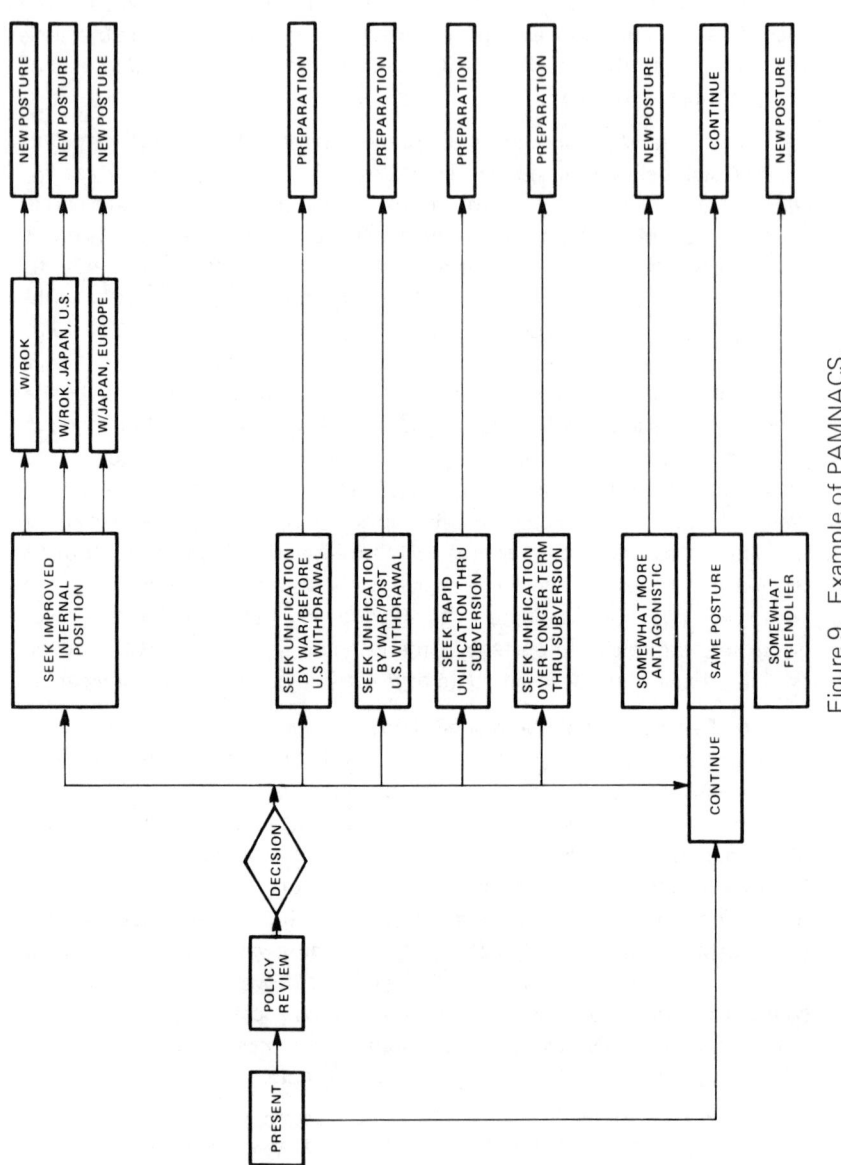

Figure 9. Example of PAMNACS.

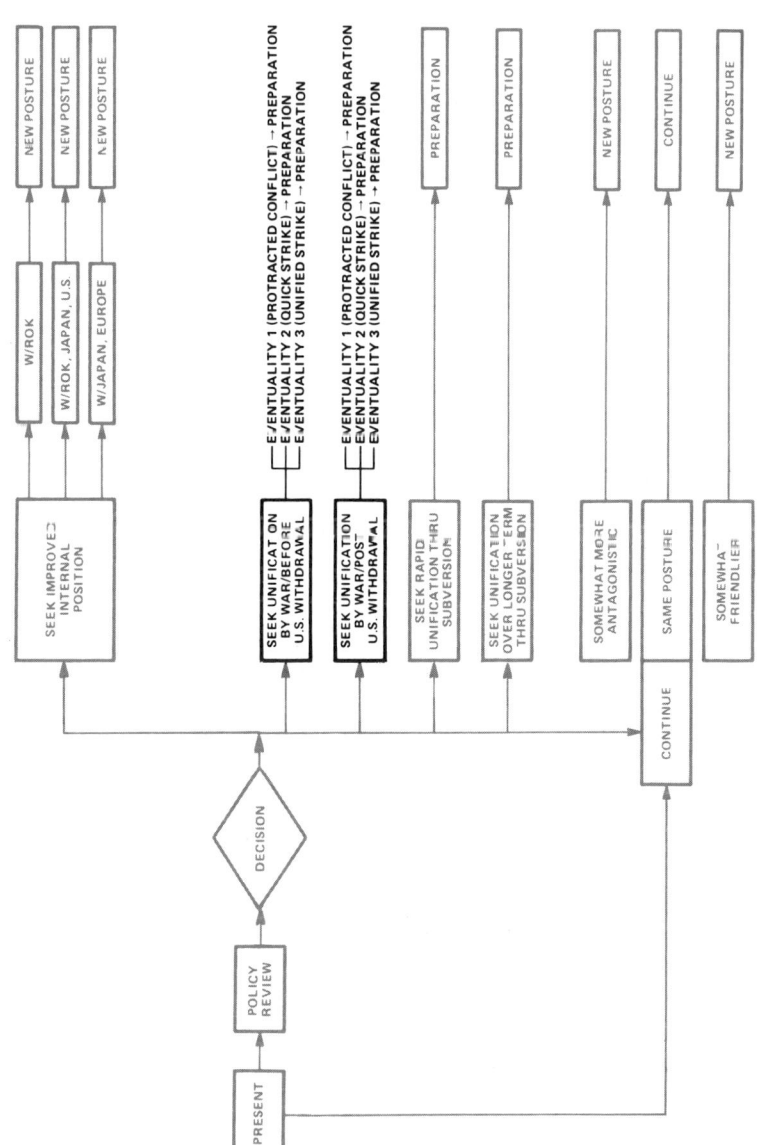

Figure 10. Second Example of PAMNACS.

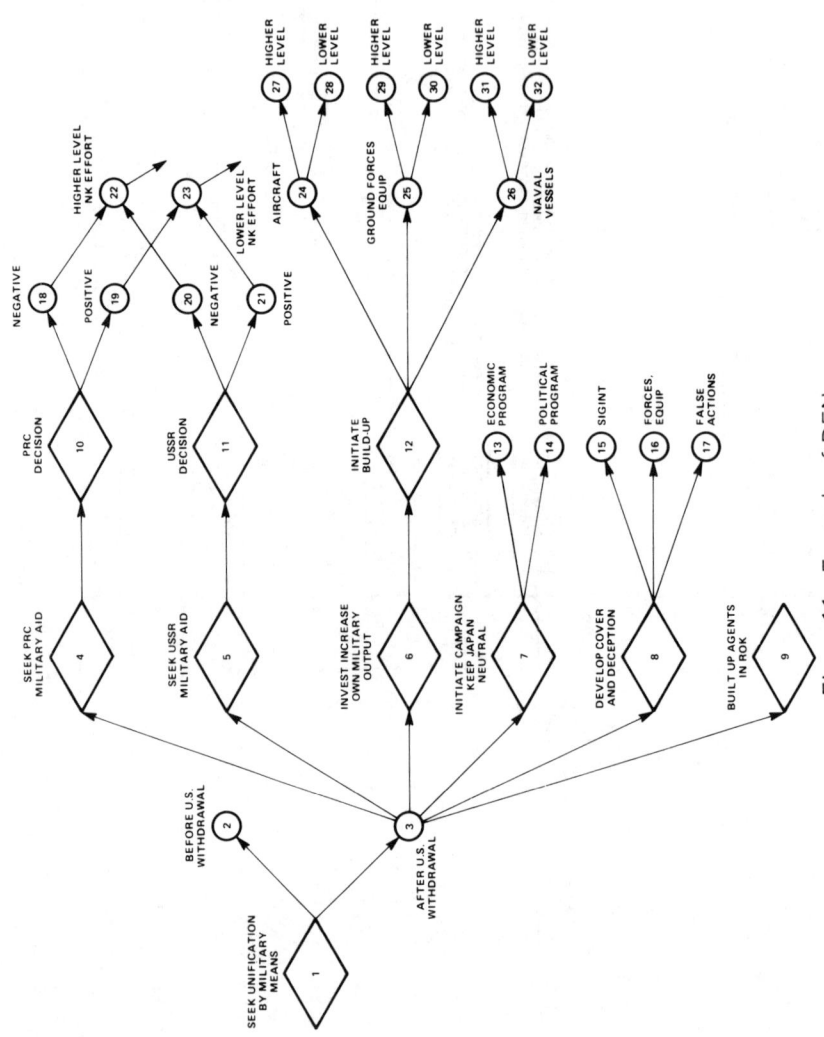

Figure 11. Example of DEN.

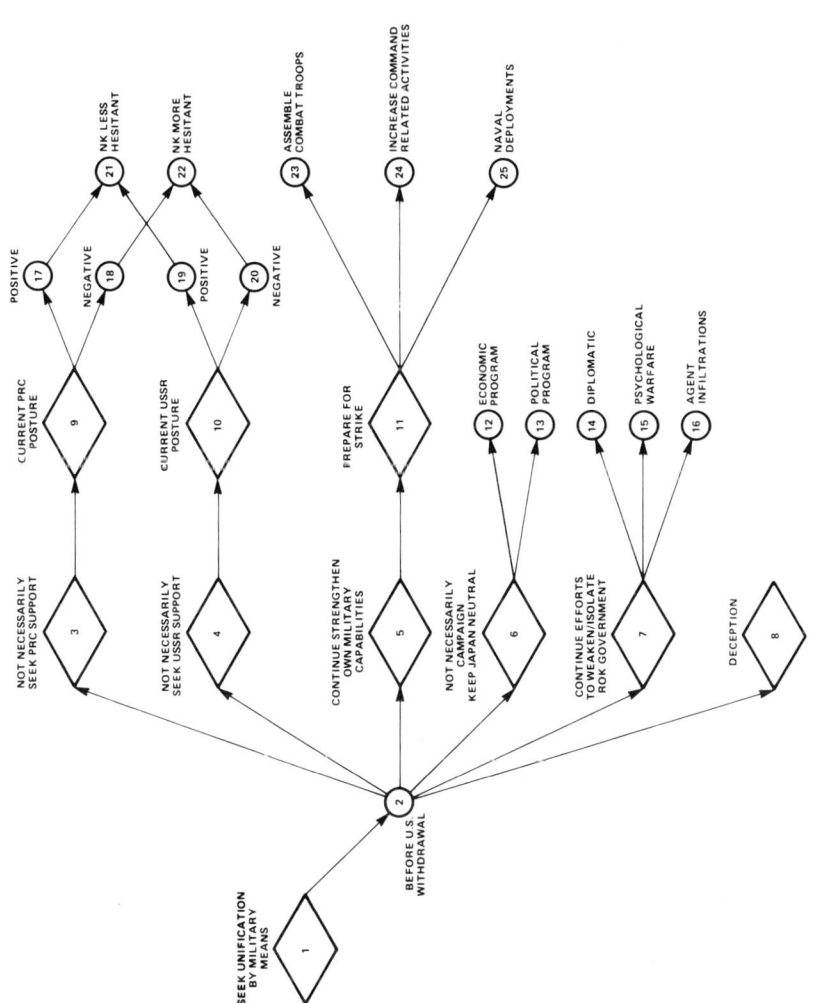

Figure 12. Second Example of DEN.

The numbers shown at the nodes of the DENs are for purposes of indexing and referencing the next type of form, the *critical event filter*.

- **Critical Event Filters (CEF).** Figure 13 shows an example of a CEF that applies to the PAMNACS. Similar examples would apply to the DENs. CEFs may be thought of as containers for holding data that are judged to *signify* that the activity modeled at given nodes in the PAMNACS and in the DENs is occurring. For example, with reference to Figure 12, Node 23, incoming data on the convergence of North Korean ground units in areas outside their garrisons could signify the assembly of combat troops and hence render that node active. In a computer-based memory, the CEFs could be considered a portion of the memory in which such input data would be stored.
- **Anomalous Event Matrix (ANEM).** Although always a goal, obviously it will not be possible to model all threat contingencies in advance. Novel situations will develop and recognizing them is one of the major challenges to strategic analysts.

It is important to be alert to those cases in which incoming data does not match well with the preestablished context. In these cases, the analyst should have learned that he must take the precaution of formulating new (or novel) threat parameters in order to detect any situational patterns developing in the data. Procedures for developing new threat parameters will generally follow those for designing threat models *per se*. The analyst should be particularly sensitive to his limitations in knowledge and expertise, and insure that appropriate technical consultation is obtained if there appear to be "abnormal" or "inexplicable" relationships in the data.

The analytic form denoted, *Anomalous Event Matrix*, is one approach to aiding in the difficult tasks of recognizing novel threat situations and of dealing with new data. The technique involves special examination of input data which does not relate satisfactorily to established models. The examination consists of developing a simple matrix in which incoming data is compared against new hypotheses; possible changes in technical capabilities; possible new policies in a foreign country; etc. The intended result is to develop new perspectives which tie the incoming data together better than the existing ones. In turn, these new perspectives can be used to modify or expand the existing models or to create new ones. This analytic process is of enormous significance in several dimensions, as discussed below.

Analysis Routine. The forms and associated modeling involve an analysis routine. The routine is ultimately based on the functional model of analysis in the threat recognition stage as described in Chapter 2. That stage consists of four major analytic steps and 17 distinct analytic procedures with incoming data. We must now relate the modeling to the routine.

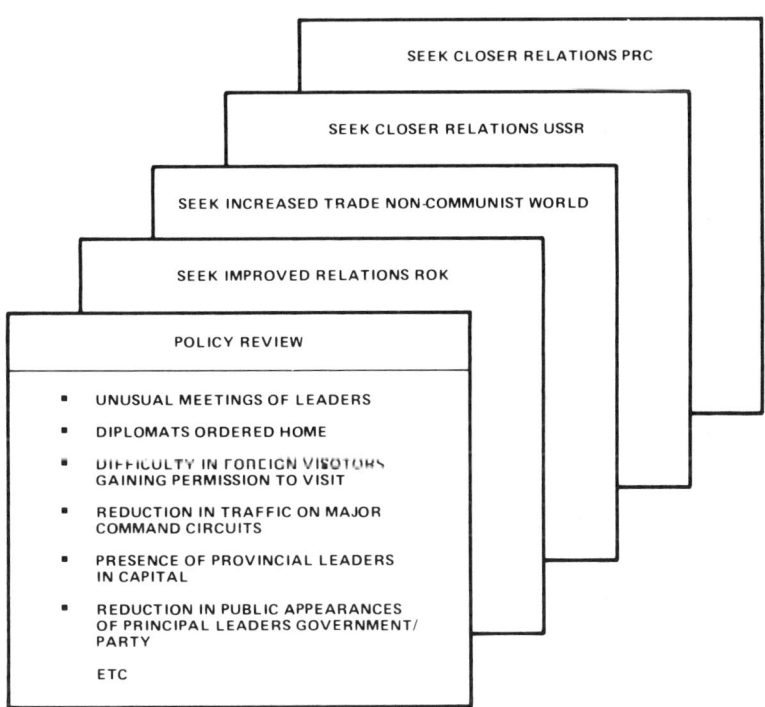

Figure 13. Example CEF.

We should begin in general terms by considering a major thrust of strategic analysis: to search for, and discover, meaning in incoming data by relating it to a context of models through the use of certain procedures. Here the analyst is embarked on the analytic adventure, a difficult mental expedition taxing his full powers and demanding that he try to overcome the human limitations that work against his analysis. In theory the analyst should not be satisfied until he can account for data, can assign it meaning. Moreover, he must not be satisfied until he has deliberately followed certain procedures, and exploited certain machine-based modeling capabilities, designed to mitigate the inevitable problems arising from his natural inferential strategies. (Although this latter point is crucial, I will allow it to remain more or less implicit in the present discussion of analytic routine but will focus sharply on it in a later commentary on signification.)

Perhaps the pervasive benefits of full analysis pursued systematically and with rigor — in short, using a routine — can best be suggested by looking at the problem of novel threats.

We said above that the analyst must be alert to those cases in which incoming data does not match well with the system of preestablished models. Conceivably these cases may signify a novel threat situation. At the very least they may signify that the present versions of PAMNACS, DENs and CEFs are too narrow to account for the perceived events and activities. Therefore, the effective assignment of meaning *cannot be accomplished until the models, and perhaps procedures as well, are expanded or modified.* Put another way: the analyst cannot effectively analyze such data all the way through the full process until he invents new structure and analytic strategies to do so. In the regenerative process of doing this — of responding to a problem of signification — the analyst may discover a new threat, a new dimension to the problem, a new indication and so on. This will create analyst-inspired change to the structure of the analytic system itself. Through the changes he makes, the analyst will tend to mitigate problems of atrophy and to refine methods of analysis. He will tend to stay more current on the problem area he is responsible for. In sum, if the interpretive impulse is facilitated, not blunted, the analyst has the chance to be, and to become increasingly, effective. The paradox, profound and subtle, is this: only by working systematically and hard to create a set of imperfect models does the analyst create the basis for improving his effectiveness. His chance to improve his effectiveness lies in his interpretive impulse which causes him to worry the new data against the imperfect models, out of a mixture of skepticism and excitement, and to make refinements. The analytic routine helps him to be thorough. The context of models or schemata helps him to see when he knows something and may be right; and when he does not and must create ways to do so. As I have said earlier, he must, because of the ways in which his mental capabilities have evolved, rely on extrasomatic memories and extrasomatic analytic support in accomplishing these goals. In Chapter 5 I will speculate in some detail on the strategic analyst as innovator.

We should consider briefly certain other aspects of the cognitive process in strategic analysis, for when the analyst follows the threat recognition routine he does much more than is expressed in the individual steps and procedures outlined in our functional model of strategic analysis. Indeed, among the unique aspects of strategic analysis are the special roles of memory and imagination. In strategic analysis there is a crucial requirement for effective memory. An analyst must recall associations and relationships, past behaviors and past events in order to understand current information as a basis for projecting the future. We remember by recourse to our information base of actually sensed events. Further, memory is colored by imagination: given that the mind simulates reality from the abstractions of sense impressions, it can equally well simulate reality by recall and fantasy, creating stories and playing imagined and remembered events back and forth through time. Obviously these powers apply to the modeling of hypothetical threat situations (as well as to projections) in the process of strategic analysis. It may seem a grimly humorous question in light of certain history, but is the analyst doing anything fundamentally different when he makes such projections as against when he daydreams? In purpose and mental direction, the answer is (or should be): yes, he is doing something very much different. But in terms of essential mental processes, he is not.

Hence we must distinguish various kinds of imaginative activity. With some wariness, we might consider *disciplined* versus *undisciplined* imagining. If one is fantasizing a pleasant and personal future situation — and all fantasizing ultimately implies the future — then one is likely to be in a kind of meandering, free-association mode. In some psychology this process is exploited for therapeutic purposes. In some literature it is modeled as part of the "stream of consciousness" tradition in an attempt to convey the sense of a human being by showing patterns and tendencies in "unstructured" modes of imagination. If there are problems of mind set, bias, wishful thinking, distortion, mirror imaging and all the rest of the catalog of pitfalls in human perception and cognition, it is good that the problems arise and are evident, both for the therapist and for the writer who is trying to dramatize a particular human personality.

But man the analyst must do directed threat modeling and projections, which of necessity must be done through "disciplined" imagination. They must be done imaginatively because, as I have said, the human being has no other way of challenging the future except by going back through the data base of remembered images, scenes, hypotheses, convictions, etc. and creatively putting together that future. The sobering aspect is that strategic analysis has as its basic thrust the Sisyphean attempt to understand the future with literal realism, with great specificity across all the information categories: who, what, when, where, how and why. It seeks under tremendous constraints to model and project what will happen both in an unimpeded sense and an influenced sense. Strategic analysis thus becomes a uniquely demanding, creative act. Hence the routine becomes essential for disciplined approaches.

When a novelist or playwright or movie director creates an idealized world full of people, events, conflict and, ultimately, meaning (even if it is the view that there is no meaning), a projection of sorts has been made. An enormously complex, sophisticated body of technique and narratology — all of it highly disciplined — has grown up around the making of novels and plays and movies. On a different level, the strategic analyst is also a narratist. There are general techniques of narratology highly germane to strategic analysis in modeling and projection.

Consider that the strategic analyst undertakes to develop scenarios. What are the techniques and stylistic features of the scenario? The ability to construct realistic and plausible scenarios presupposes not only a comprehensive knowledge of an area and its possible trends, but also demands varying degrees of relatively specialized literary skills. The basic problem is to take the essential ingredients of a hypothetical situation and order them into a coherent, realistic sequence of events. Existing scenarios will require periodic refinements to lend them greater impact and credibility on the basis of new data. This process is also, of course, part of the dynamic of signification.

The analytic forms described above are designed for flexibility in scenario development. One of the basic advantages of scenarios lies in their diversity. They can vary considerably, both in subject matter and form. However, the essential nature of any scenario is that the situation described is hypothetical. By definition, a scenario is a depiction of a hypothetical series of events that could produce some envisaged situation. Thus the analyst's basic problem is to establish a coherent pattern of events within which the projected situation could plausibly occur. In doing so, he is establishing structures of rationale. These in turn are bound up with signification.

Much has been written about achieving plausibility in writing and in other art forms that are largely imaginative (or hypothetical). In its broadest dimensions, the problem is that of using all the elements of composition effectively: mechanical elements such as narration and description; and conceptual elements such as plot and unity. This challenge entails disciplined approaches. However, the great majority of scenarios fall within relatively narrow limits as to the mechanical and conceptual problems they pose and thereby require rather straightforward routines. One reason is that scenarios, though they need not do so, are generally intended to serve as broad outlines of likely processes. Their characteristics include the following:

- Scenarios, unlike dramatic fiction and informal essays, seldom tend to digress. Frequently they will be pure plot in the way history sometimes is reported: event A leads causally to event B, which in turn leads directly to event C, and so on until some final situation or set of prevailing conditions is reached.
- Scenarios frequently consist primarily of visual diagrams designed to condense action into its largest movements. That is, they are intended to depict long-range views rather than highly detailed close-ups of actions and individual actors, though the latter forms are certainly not precluded. These types of scenarios often will telescope the events of several weeks, months or even years into a relatively short space.
- Scenarios are often deliberately employed to incorporate and integrate numerous factors in a situation, including military and political, sociological and economic, and psychological and cultural influences and processes. Thus they may include more frames of reference than ordinarily are found in imaginative writing and analysis.

From these characteristics certain problems requiring special attention by the analyst become apparent. One of these is to make sure that at all stages the coherence of the plot or sequence of events is immediately visible: that an outcome is believed possible *because* certain earlier events or conditions are believed capable of bringing it about. In turn, it should be clear how this outcome influences other stages in the course of events being described. Again, rationale structures are being developed.

Table 2, Chapter 2, shows the functional outline of the threat recognition stage of analysis. In reviewing the table, the reader will see quickly that there is a close correspondence between the modeling activity — PAMNACS, DENs, CEFs and ANEMs — and the steps and procedures in the analytic functions. Largely through the modeling, the routine is implemented: the indicators are matched against the various models, the correlations identified, and the models as a set compared in terms of their relative levels of activity. Novel threat analysis is conducted, with the steps and procedures implemented through the ANEM. Hence the modeling is a means of facilitating the process. Backlog and relationality measures enable us to monitor the extent of these activities.

Signification. In the threat recognition stage the analyst, according to a chosen rationale, assigns meaning to the incoming data in association with a context of models or artificial schemata. Having said this, we must reflect on meaning and context. I have said that meaning constitutes a change in a living system's processes elicited by an information input. For the individual strategic analyst and for a group of analysts pursuing the strategic mission, changes that indicate meaning has occurred are any and all changes in the perspective on the likelihood of strategically important situations. Incoming data must have more meaning or less meaning — but definitely *some* degree of meaning — in relation to what it is thought to tell us with respect to possible situations for which we must be alert. The information may appear insignificant, mildly interesting or highly indicative. Moreover, this meaning is visible through changes *in* or *to* the contextual models used in analysis.

The anthropologist, Edward T. Hall, has written importantly about the relationship of context to meaning. Hall explores richly the distinction between high context and low context modes of thought and expression. High context modes have a considerable amount of rationale and perspective already built in and available for the interpretation of incoming data and the development of outlooks on states of reality. One can think of such modes as involving established models of reality and systems of rationale (and, of course, biases as well) which create a strong context for the assignment of meaning to input data, all hopefully in the interests of perceiving and interpreting reality for purposes of developing policies and strategies of enlightened self-interest and survival. High context modes are intended to incorporate painfully learned lessons of experience which have some proven reliance as continuing guidelines for behavior. They may be stores of collective wisdom obtained over long periods and intended to keep us from repeating tragic history. (They are also deeply involved with culture and cultural identity and integrity.) High context systems have the advantage of simplifying the cognitive functions of analysis and decision. They tend to ease problems of information overload and stress.

High context modes of expression tend formally to incorporate high levels of explicit signification, particularly in terms of capturing rationale. A simple but penetrating example is given by Hall in his description of how the Hopi Indians use their high context language in discussing the weather. Hall begins by commenting on spoken and written contexts:

> . . . context . . . carries varying proportions of the meaning. Without context, the code is incomplete since it encompasses only part of the message. This should become clear if one remembers that the spoken language is an abstraction of an event that happened, might have happened, or is being planned. As any writer knows, an event is usually infinitely more complex and rich than the language used to describe it. Moreover, the writing system is an abstraction of the spoken system. . . .in the process of abstracting, as contrasted with measuring, people take in some things and unconsciously ignore others. This is what intelligence is: paying attention to the right things. The linear quality of a language inevitably results in accentuating some things at the expense of others. Two languages provide interesting contrasts. In English, when a man says, 'It rained last night,' there is no way of knowing how he arrived at the conclusion, or if he is even telling the truth, whereas a Hopi cannot talk about the rain at all without signifying the nature of his relatedness to the event — first-hand experience, inference, or hearsay.

The danger in a high context system is atrophy. It is a poor wager indeed to suppose that reality will not eventually change in such ways as to render at least portions of high context systems obsolete. I need not belabor the point that the discussions above of certain pitfalls in our natural inferential strategies, especially those involving the use of knowledge structures or schemata (including artificial schemata), apply here.

Low context modes can intensify the interpretive effort, since the human must invent new context in order to assign meaning and hence understand and explain perceived reality. This requirement brings with it all the problems that high context modes tend to mitigate, particularly those of information overload. But it also has the advantage of causing us to look at new data from new perspectives. It tends to help us avoid failing to recognize novel situations because we are too wedded to existing context which may have gone out of date.

Obviously the ideal strategy in strategic analysis is to achieve the appropriate balance between high and low context analysis or some condition of oscillation between the extremes. It is evident that the system of models — PAMNACS, DENs, CEFs and ANEMs — represents a high context interpretive mode. The models establish a substantial context within which to assign meaning to new data. But we have noted that the crucial factor in strategic analysis is the force of the interpretive impulse which causes the analyst to challenge the high context models when he cannot assign meaning satisfactorily within their current forms. As an analyst begins any given sequence of analysis in the threat recognition stage, he may experience either of these cases:

- **Case 1:** He finds that new input data may be readily associated with a node in an existing model.
- **Case 2:** He finds that the data is not readily associated. This condition then requires him at minimum to modify the existing framework or create a new framework to account for the data. It may even require him to create an entirely new model (e.g., a new major national course of action in the PAMNACS, followed by a new DEN and associated CEFs).

In Case 1, the primary record of signification is held in the CEF. Recall: the CEF is not only a model node (and a carefully chosen one, as we discussed in the earlier comments on the art of scenario building): the CEF is also a kind of signification reference point; it is a container in which we store input data which we associate with real activity similar to the activity modeled at that node. The analyst, by associating the data with that particular node, is saying that one of the conditions in a threat situation he believes he has effectively modeled is occurring.

The *full* rationale behind the assignment of meaning — that is, behind the signification of the data — lies in the assumptions and logic that went into the structure of the model itself: assumptions about the time frames required for certain events; the preparations needed; and any and all other modeled conditions believed necessary for the situation to occur. Here I must ask for the reader's patience temporarily. I cannot deal fully with the issue of the assumptions in models until I duscuss the projection stage of analysis. The place to deal with the assumptions behind models is in the discussion of the analysis of conditions leading to threats, and that analysis is bound up in the process of making projections. Although we are describing strategic analysis as a linear process, it often becomes a cyclical, iterative process in real operations. Developing threat models is done in conjunction with other kinds of analysis performed in the projection stage.

In Case 2, the primary record of meaning lies both in changes made *to* the models and changes made *in* the models. The incoming information, because it has not appeared to have adequate meaning within the context created by the existing forms of the models, has caused the analyst to search for meaning through the creation of new context. This begins as low context analysis whose thrust is to create a new high context useful for future analysis.

I will interrupt the discussion of signification at this point, returning to it when we consider analysis in the projection stage, where we can more appropriately explore further dimensions of the problem.

Cognitive Technology Implications. I have argued that specialized machine support to the analyst is essential. I have said that this is so because there must be extrasomatic memory and analysis aids to mitigate a variety of problems which constrain human analysis. Later in this chapter I describe what I believe to be representative software requirements for such support: programming needs, display designs, major man-machine interface requirements and other traditional design considerations. What we should consider in the present section are some important general implications about the basic nature of such machine support to the strategic analyst.

One consideration has an overriding importance. It is speed of operation. The human mind can operate with enormous speed in processing and analyzing data. It may worry a problem for long periods of time, creating and reviewing options, imagining consequences of action and contemplating problems from many sides. *But as it does such analysis — in the analytical acts themselves — it naturally operates at great speeds.* Successful cognitive technology must allow the analyst to continue to operate at adequate speeds. Ultimately there are two words to describe the problems that otherwise ensue: *delay* and *interruption*. If there is interruption of the all out, speedy processes of cognition in hard analysis, whether the processes be lightning-like associations, a series of associations made over a few seconds, or a general assault for a lengthy session on a puzzling problem, such as the significance of an anomaly, the result can be devastating. Anyone's everyday experience verifies again and again the negative effects of interruption on difficult analysis. No particular expertise in human factors is needed to realize that out of the many reasons why analysts may not depend on a given computer-based aid, one of the primary ones is that it fails to meet the need for sustained analysis at appropriate speeds. Certainly one of the diagnostic goals in monitoring the backlog and relationality variables is to discover such problems anywhere in the analytic process.

There is a serious question about the limits in strategic analysis on the effectiveness of *any* approaches using extrasomatic memories and analytic aids embodied in books, reports, memos, etc. It is, of course, unrealistic to imagine that we will cease to depend substantially on these over the near term. Nor can analysis be conducted entirely in a visual mode using computer-driven graphic displays primarily comprised of geometric configurations and alphanumerics. But the simple fact remains that books, reports, memos, etc. are harder and slower to use in virtually all aspects of analysis: data review, data correlation, interpolation, integration, etc. Because of this, they are more likely to reside in container drawers, ignored and allowed to go out of date. And because of these and other problems, they will not provide an adequate record of analytic operations and processes.

Tolerance of delays and interruptions is to some extent dependent on the specific capacities for such in each individual. And, of course, sometimes delays and interruptions may ironically prove beneficial. Furthermore, certainly the written word in paper storage has uniquely beneficial properties as a memory and as a vehicle for sophisticated analysis. Each media has its virtues. But none of them is even remotely as important as the overriding point. It would be foolish indeed not to understand that speed of analytic operations made possible by computer-based support systems, together with the secure memory these systems afford, is directly related to enormous possibilities of increasingly sophisticated and realistic human interpretation of reality. Throughout the present research there is no more important point. Nor is there a clearer imperative than in the fact that all too few manhours doing such analysis on significant scales have ever been expended.

There is no necessary conflict between strategic analyst and machine. The machine not only need not constrain the analytic capability of the analysts; it *should* extend it, increase the productivity of the energy of the human analyst, and enable more, faster and less error-prone analysis.

In terms of the models and modeling in the threat recognition stage, a primary design goal for computer-based support becomes clear: the calling up, comparing and modifying of models and portions of the models must be capable of being performed quickly and displayed succinctly and effectively. As I shall discuss below, there is remarkable potential in the use of color displays and other devices for such purposes. The analyst must be able to manipulate the extrasomatic schemata readily and speedily to sustain the momentum of the analytic processes in his own mind. In doing so, he will create a record for *post mortems* that also must be reviewable with appropriate speed as the need arises.

Projection Stage

The strategic analyst now faces his greatest challenge: the projection of situations. As I have said, he will not be able to anticipate final victories. Indeed, we have discussed in some detail the fundamental limits on projections by human beings.

But what prospects are there for improvement in the projection of strategic situations? That is the important question. The need to make projections sufficiently trustworthy as to inform enlightened policy simply must continue to become increasingly a recognized priority. Yet until a considerable effort to do such projections is mounted, an effort involving multiple types of information, computer-based memories and analytic aids, substantial communities of analysts and thinkers operating with continuity and with constant self-review, and with R&D response to shortfalls, we will really have no firm basis for assessing prospects. However, can there really be any acceptable alternative to the determination to improve our capabilities?

In the present discussion, the intent is not to recommend a single best way of approaching projections. Obviously there are projection methodologies yet to be invented which will better the current inventory. We will discuss a general method believed promising as a basis for evolving more refined methods. It is an eclectic approach, an admixture of various methods. There is no intent to provide a survey of various projection techniques, nor is there an implication that what is included is necessarily more suitable than, or implies a reason for excluding, other available techniques. Such qualifications become less important when we recall that the focus is not really on any current set of methods nearly so much as on a man-machine system whereby communities of analysts may readily and heuristically evolve improved projection methods. In short, the projection system described is a start.

Analysis Forms. The making of projections presupposes a set of forms used in conjunction with those in the threat recognition stage. We can conceive of cases in which the analyst, having reviewed incoming data in the threat recognition stage and detected early signs of a potential situation of strategic importance, then proceeds to projecting outcomes. It is at this point that the requirement for additional forms or schemata arises. In discussing these forms we must simultaneously touch on aspects of analytic routine; there is also a separate section on routine *per se* following the discussion of forms.

- **Translation of Threat Models Into Projection Format.** The first form is referred to as the *projection format*. As Thomas Belden has pointed out, the strategic analyst, whether developing unimpeded or influenced projections, ultimately has the obligation to respond to the basic information categories — who, what, when, where, how and why. Each presents a separate analysis challenge or obligation. At the same time they must be combined into strategic judgments. This implies that a grand projection methodology must be a synthesis of several submethodologies dealing with various problems associated with the different categories. Similarly, the measures of analytic performance must penetrate each category. In discussing the projection stage in the strategic analysis model, I established as the first procedure the translation of individual threat models used in the threat recognition stage into a projection format. The format, first adopted by Belden, is this:

(1) *Who* (or what)
(2) (Could do) *what*
(3) (To) *whom* (or what)
(4) *Where*
(5) *How*
(6) *When*
(7) *Why* (both to what end and/or because x conditions apply).

Of course this form applies whether we are considering a single statement of a threat or an extended description of one.

The information categories, combined into the English sentence, form a penetrating and elegant way to structure the problem of strategic projections. The proper focus on strategic projections should be on the essential output to decision makers and other users; and, as simply and directly as possible, on each of the major areas of challenge to the analyst in developing that output. It is useful to consider briefly the obligations of the strategic analyst in developing projections within each information category.

First Analysis Obligation — Who and What. Who and what are explicitly contained in sentences formulating a basic strategic concern:

Who might do *what* to *whom*? and
What might cause *what* to happen to *whom*?

For analysts, the problem of *who* and *what* in reference to initiators or catalysts of threat enactment essentially reduces to identifying countries, political and military groups, leaders/decision makers, and economic, psychological and cultural forces which can enact threats or cause threats to be enacted.

The problem of *what* as denoting a hostile action — nationalize industries, attack property and citizens, etc. — and the problem of *whom* as the victim of the hostile action, are closely related. Any generalized list

of types of strategically important interests and assets and the potential threats toward them would include items such as these:
- Attacks against diplomatic, military, and corporate personnel, facilities and equipment located outside the U.S.
- Hostile actions directed against other U.S. citizens located outside the U.S.
- Seizure of U.S. overseas commercial investments.
- Denial to U.S. of use of lines of communications.
- Withdrawal of political support in world and regional organizations for U.S.
- Economic activity directed against the U.S., such as formation of cartels to deny combinations of vital resources.
- Attacks against U.S. treaty allies.

As noted, a major principle implicit here is that the proper focus of strategic analysis is primarily (though not exclusively) on threats to friendly interests and assets. Identifying potential initiators and catalysts of threats, and identifying and prioritizing the assets and interests threatened, can be considered the first obligation of the strategic analyst.

Second Analysis Obligation — Why. The second issue confronting the analyst is: Why. This can be stated as:

Who might do what to whom *because* —

This need to address why is perhaps the most fundamental and certainly the most challenging obligation of strategic analysts. The overall problem of credibility, including credibility with decision makers, obviously is linked to establishing the plausibility of potential situations, particularly adverse ones. In the most cosmic sense, theories of what determines the course of events range from the view of Carlyle that the activities of leaders are the predominant factor; to the outlook of Tolstoy which views history as an accumulation of accidental events with the element of chance more important in determining its course than the decisions of leaders. Both theories can be useful; either may dominate, depending on the situation. More important to the problem of method and procedure is that the analyst must have succinct, up-to-date information based on:
- The goals, predispositions and characteristic behavior patterns of key decision makers
- Decisive military, economic, political and cultural forces.

Further these must be tied in with threats to strategic interests and assets.

Third Analysis Obligation — When, How and Where. The third obligation is to consider *when*, *how* and *where* a threat might be enacted. Each problem presents a different challenge to analysts.

For the problem of *when*, it is necessary to attempt to take into account all the necessary processes that must occur. It was noted that the strategic analyst can focus on a finite (and, of course, prioritized) set of threats. For the problem of *when*, it is important that the analyst attempt to identify the essential processes — military, political, economic, psychological — that must unfold in the enactment of given situations, and then estimate the minimum or worst case as well as other potential time lines involved in the unfolding of these processes. Important items in threat enactment could include such factors as the extent of emergency powers of a government, its command and control capabilities, terrain, and the nature of U.S. assets threatened.

The problem of *how* requires that the analyst consider factors such as the means of threat enactment at the disposal of foreign decision makers and the economic, political or other forces that might combine to produce the enactment of a threat. The objective is to estimate the relative likelihoods of occurrence of plausible processes of threat enactment. Factors such as the size and nature of police/military forces of the hostile government and the political constraints on that government would be significant.

As regards *where* — the problem of projecting the geography of threat enactment — the geography of friendly assets and interests clearly is a key factor. To varying degrees, political, military and environmental factors will also be important.

Table 6 provides a very simplified, hypothetical example of the translation of a threat model into the predictive format. Such formats could be used for various strategic situations and both types of projection, and could accommodate considerably more detail and specificity.

- **Specificity Ratings.** A prototype Specificity Rating Scale has been developed for use in comparing the specificities of models, predictions and forecasts across the information categories. Table 7 exemplifies the form, showing the thrust of the approach. Of course, refinements would be developed through operational use of the scale. In computer-based operations, specificity levels could be coded with color and displayed on color graphics. In turn, models used in threat recognition and projection (see below for examples of the latter) could be displayed with the color codes as a means of rapidly and effectively indicating comparative specificities.

- **Recourse to Basic Principles Underlying Projections.** We have discussed the imperfection of projections made by human beings. Yet in his obligation to make projections, the strategic analyst must assume that there are *some* reliable rules or principles of change to which he has

Table 6. Example Use of Information Categories With Threat Models

1. *WHO* (or WHAT): North Korean combat forces

2. (Could do) *WHAT*: could be directed to launch surprise attack from present positions

3. (to) *WHOM* (or WHAT): against the ROK, and the U.S. military forces stationed in ROK

4. *WHERE*: by attacking across the DMZ and into ROK via various routes

5. *HOW*: by attacking without mobilization or forward movement of reserve forces prior to attack, and by using units now deployed near the DMZ; by flying limited interdiction strikes without a major forward deployment of aircraft; and by conducting limited naval operations shortly after ground forces cross the DMZ

6. *WHEN*: within one week

7. *WHY* (or because X conditions apply)

   a. Assume North Korea's long-range goal remains the unification of the entire Korean Peninsula under the North's regime

   b. North Korea perceives an anarchic situation developing in ROK via the following: uncontrolled student demonstrations; a *coup d'etat*; an economic collapse; an ROK government leadership crisis

   c. North Korea perceives that the ROK is losing its ability to offer determined resistance

Quantitative Value	Information Category						
	Who[1] (Identity of U.S. Asset or interest)	Who[2] (Identity of Foreign Actors)	What (Identification of Threat)	When (Time of Crisis Accurate to the :)	Where (Location of Crisis)	How (Description of Implementation of Threat)	Why
S_{max}	10	10	10	10	10	10	10
INCREASE SPECIFICITY VALUE	Unique identity of asset	Unique identity of foreign actors	Specific threat: Confiscate Kidnap Armored Attack	Minute Hour	Unique location identity (e.g., individual structure location)	Exact form, direction, speed of "tactical" threat (e.g., ground, air, sea, etc.)	Specific causes (e.g., economic crisis brought on by crop failure, etc."
	Group or type of asset	Group or organization of foreign actors	General threat: Increase Tension etc.	Day Week	Area within region (e.g. area within population center)	General form of threat (e.g., "military", "economic", etc.)	General causes (e.g., facism, etc.)
	Sector identity of asset	Country of foreign actors		Month	Region (e.g., population center)		
	U.S. asset or Allies Interests			Year	Latitude/longitude		
	etc.				Country		
					Section of Hemisphere		
					Hemisphere		

Table 7. Example Specificity Rating Scale.

recourse in attempting to understand the future of social systems, their processes, technologies, behavior patterns and so on. In his very useful book, *An Incomplete Guide to the Future*, Willis Harman has formulated a wise approach to making projections. Harman's focus is not on strategic analysis as we are examining it, but portions of his thinking apply to our discussion. Harman reminds us that although there are few clear-cut ways in which the behavior of social systems are dependable, one may usefully distinguish several general principles to consider when concerned with the dynamics of complex systems. These are *continuity, self-consistency, similarities among systems, cause-effect relationships, holistic trending* and *goal-seeking*. Ultimately the various methods of projection, from the highly qualitative to the essentially quantitative, are based on various mixes of these principles. It is my contention that these principles should be considered systematically by strategic analysts making projections.

What, specifically, do the principles encompass? Here are brief definitions largely adopted from Harman:

Continuity. Large systems more often than not *flow* into new states rather than changing discontinuously. It is dangerous in strategic projections not to realize that even in times of military conflict and other conflict situations, large portions of the social structure, government organization, monetary conditions and technology of a nation may prevail. Indeed, in making projections we should strongly consider past experience. Continuity is a useful principle to stress in such projections as those involving economic patterns and military capabilities and postures.

Self-Consistency. A second guiding principle is that such systems often are internally self-consistent. For example, a given national sector ordinarily will function in ways consistent with the activities in other sectors. Large-scale industries in consumer luxury items, for instance, are unlikely to develop in a small, relatively poor nation whose government is pursuing a rapid military buildup. The principle of self-consistency is crucial to the scenario technique. As Harman notes, the purpose of developing scenarios about the future is to make certain that however we postulate future conditions we systematically examine them for coherency and plausibility.

Similarities Among Systems. Systems in different places and times will show definite similarities. For example, military juntas will tend to act somewhat alike under similar circumstances. Harman notes that this principle is used in what he calls "anthropological" approaches to studying the future and in cross-cultural comparisons.

Cause-Effect Relationships. Some of the dynamics of nations, organizations and groups may be partially understood in terms of cause and effect relationships. For example, in the analysis of economic dynamics scarcity is usually interpreted as likely to bring about rising prices. This principle is basic to "cross impact analysis," a method of exploring the future by modeling the imagined cross impact of various events on one another.

Holistic Trending. In evolving and changing, systems "behave like integrated organic wholes. They have to be perceived in their entirety." An important implication is that human analyses and judgments about the future states of a system are crucial. No matter what the inventory of machine-based projection algorithms, interpretation and signification must finally involve the human analyst. Projection is an art.

Goal Seeking. Systems such as nations, organizations and groups have goals. These goals may be more or less "conscious" and enunciated. (Certainly an analyst may observe a system and come to believe that there is a distinction between the real goals and the announced goals.) But the major point, as Harman states, is that "change is not aimless, however obscured the goal."

- **Use of Forms Which Systematically Employ the Principles of Projection to Reduce Bias.** The key idea is that the six principles should be used systematically. Projection methodologies should be designed to foster the application of analytic energies in systematic considerations of possible change. There are a number of potential approaches using the principles. These may primarily involve quantitative data on economic and political indicators; and hard data on such changes as military technology developments. Or methods might be used which attempt to project qualitative aspects such as shifts in cultural perspectives (such as the resurgence of traditional values). Methods combining these separate approaches might be used which are primarily holistic.

The difficulty, of course, lies in avoiding bias. I have stressed the difficulty in gaining objective views of the future. Certainly the analyst runs a high risk of bias when he relies too heavily on one projection technique. As Harman points out, methods that rely essentially on numerical data are subject to bias toward input information that can be readily quantified. Similarly, methodologies which focus narrowly on theoretical issues and rational behavior may lead to analytic failures because irrational and unconscious forces are insufficiently accounted for.

An obvious strategy of the strategic analyst in mitigating bias is to use several approaches to projection. The problem will lie in judging which results to stress in the various approaches and which to downgrade in importance for given analysis problems. One tactic is to look for points of similarity and conflict in the results and pursue the latter as needed.

More basically, the strategic analyst would be prudent to consider *several* plausible future situations, not a single one, given the perennial uncertainty about the future. In the discussion of threat recognition, this strategy was apparent in the construction of models of alternative threats. The principles of forecasting can be used in a variety of ways to delineate a set of alternative paths to the future. In my judgment, little has been done to develop analytic techniques which deliberately use all the principles. The following are a few promising ones, some described by Harman and some developed in the present research, and all highly compatible with computer and computer display operations.

Technique 1 — Sector Comparison. Described by Harman, this form can be used with the threat models and with several projection techniques discussed below. Sector comparison is an analytic technique for examining the plausibility of various projected situations, given differing combinations of economic, military, technological, political and other conditions. The strategic analyst would apply the *sector comparison* technique as follows.

1) In the case of North Korea, for instance, consider a number of sectors of that nation, such as:
    E = economic performance
    S = economic structure
    M = military strength
    R = military readiness
    B = outlook of decision makers
    A = relations with key nations
    I = political control exercised by current regime
2) For each sector, define alternative conditions that together span a reasonable range of possibilities. The range of possibilities may be as extensive as the analyst judges useful. Table 8 gives simplified examples. The question mark over the last column indicates that continuous review and search for additional categories of importance must occur.
3) Examine plausible sector combinations such as $E_2 - S_1 - B_3$ etc. Use projection principles such as internal self-consistency, holistic trending, cause and effect, etc., to analyze plausibilities among such sector combinations.

Technique 2 — Probable Futures Mapping. In using projection principles to analyze the plausibility of various sector combinations, an analytic technique has been developed in the present research and designated, *probable futures mapping*. The approach is based on the fundamental concept that over a specified period of time there is a definite range over

E	B	M	I	?
E_1 Widespread prosperity	B_1 Near-term takeover of ROK/expulsion of UN forces is considered plausible	M_1 Three-to-one advantage on ground over forces in ROK; AF superiority	I_1 Current regime firmly in control	
E_2 Prosperity restricted to certain groups	B_2 Expectation of takeover of ROK by mid 1980's	M_2 Superiority in AF; rough equality in ground forces	I_2 Current regime in control of most of country	
E_3 Equilibrium, slow growth, general public satisfaction	B_3 Acceptance of ROK as reality over indefinite future period	M_3 Inferiority in forces but good defenses	I_3 Current regime in control of key urban centers only	
E_4 Recession, general dissatisfaction	B_4 Focus on internal North Korean economic and political problems	M_4 Questionable defense	I_4 Current regime in control of capital and few other areas	
E_5 Economic depression	B_5 Non-alignment stance	M_5 Military forces in poor status	I_5 Current regime extremely vulnerable to opposition	

Table 8. Examples of Sectors.

THE ART OF STRATEGIC ANALYSIS 115

which sector combinations can vary. For example, it is unlikely that a nation would go from a classic depression to substantial recovery in one year. The probable futures mapping would be done in the following steps.

1) **Detailed Probable Futures Mapping for Each Sector.** Beginning with a description of the present in terms of various sectors and their possible ranges of conditions, the analyst then extrapolates into the future and analyzes, for each sector and for every time period, which conditions are probable. In doing so he has recourse to the principles of projection. Typical single sector mappings are shown in Figures 14-17 with respect to military (M), economic (E), internal security (I), and leadership outlook (B) sectors. The analyst has decided that over specific periods of time the economic, military, internal security and leadership outlook conditions *could* change according to various and varying schedules.

2) **Synthesis of Single Sector Mapping.** The maps developed in Step 1 are now combined into a *probable futures map*, as exemplified in Figure 18. The analyst has decided on a definite range of possible combinations of conditions over several time increments. This technique is intended to help the analyst focus on the probable future and not waste his energies on too wide a range of possibilities. The arrows on the map indicate unique conditions for given combinations. As we shall discuss below, such prompting aids as the arrows anticipate a set of computer-based graphic displays.

3) **Sector Analysis.** To this point we have been concerned with possible conditions, but not explicitly with possible threats. The analyst now uses the results of the sector comparison technique to review all the *threats* of interest. The result of this step is to associate a time frame and sector combination with each threat. In this way the analysis of possible and probable conditions is related to the threat modeling performed previously in the threat recognition stage.

4) **Identification of "Hidden" Threats.** This brings us to a very important point. The mapping which results from Step 3 may display various probable futures described by unique sector combinations *not all of which have a previously imagined threat associated with them.* The analyst must study each of these probable sector combinations and either identify a new threat which falls into the sector combination and had not been analyzed before; or satisfy himself that no threat exists for that

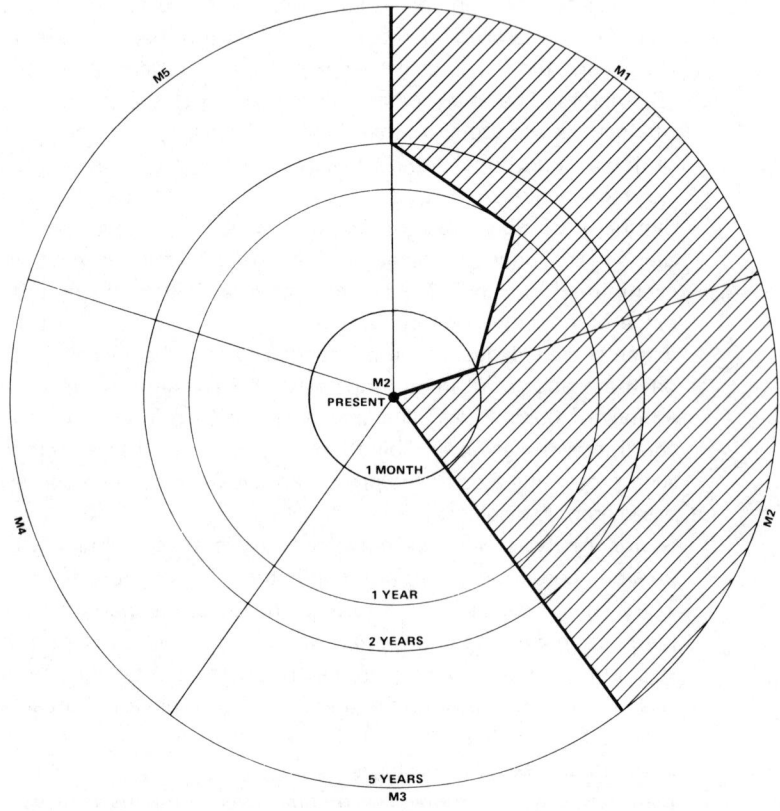

Figure 14. Single-Sector Map — Military.

117

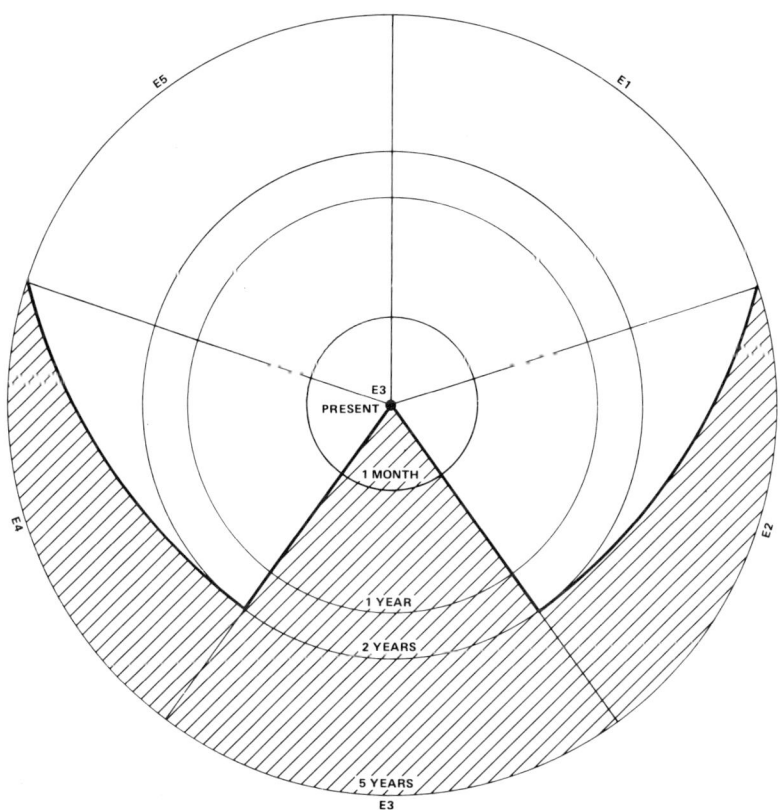

Figure 15. Single-Sector Map — Economic.

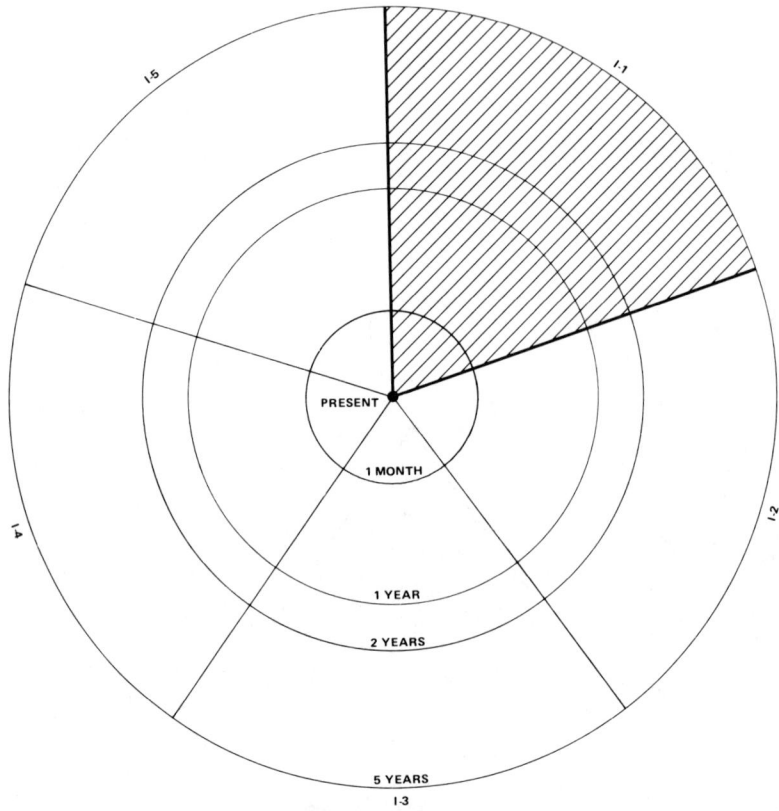

Figure 16. Single-Sector Map — Internal Security.

119

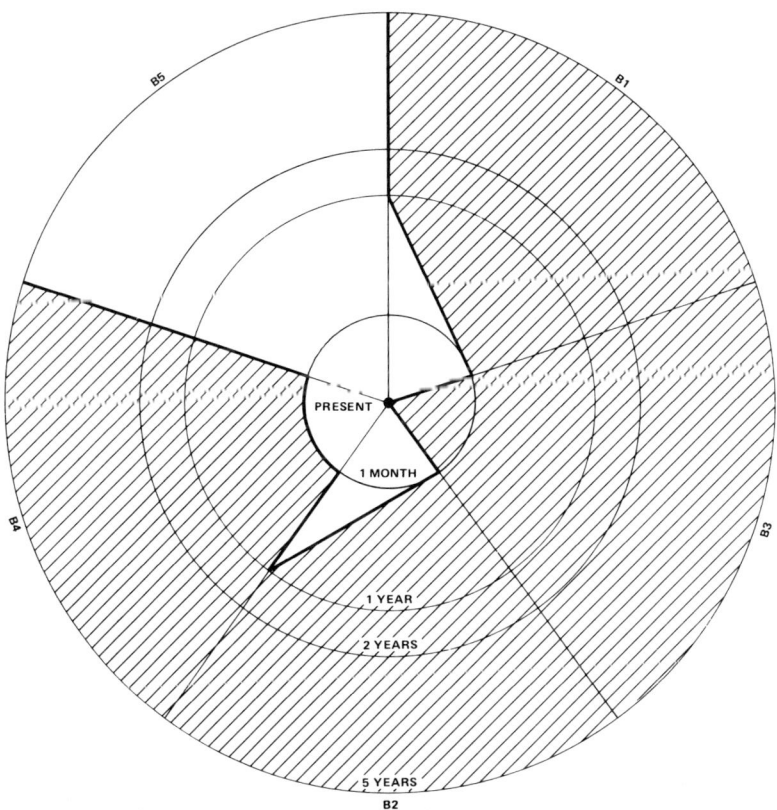

Figure 17. Single-Sector Map — Leadership Outlook.

sector combination. This step aids the analyst in discovering previously undefined threats within the scope of a probable future.

5) **Plotting the Unimpeded Threat Line.** With the probable futures map completed, and possible threats located, the analyst has the task of plotting the unimpeded threat line — the path into the future that the subject country or other strategic entity will most likely take, given no influential change in present day policies and activities of friendly decision makers and forces.

6) **Assessing Impact on Unimpeded Threat Line.** Finally, given the unimpeded threat line is not pointing to desirable situations, the probable futures mapping technique can serve as a framework for reviewing how changes may be instigated to alter the unimpeded threat line (and thereby create an influenced projection). Obviously this step may involve a link with the decision maker.

Technique 3 — Employment of General and Opposing Projection Methodologies. In order to help mitigate bias and consider potential situations in a realistic fashion, the strategic analyst can employ two general and opposing projection methodologies denoted, *present extrapolation* and *future-backwards*. These can be used with the other techniques described above.

1) The *present extrapolation* method of projection extends events from present conditions to future situations. The descriptions of the time-phased events that comprise these extrapolations may tend to be very specific in their details in periods close to the present and become progressively more general and vague further out in time.

2) The *future-backwards* method begins with a precise description of a future situation, for example, a specific threat. The analyst then works backwards in time along what he considers necessary paths. Hopefully these paths will uncover required preconditions for the future situation. Use of the two methods together can be valuable since logical discontinuities in a projection may become more apparent through a comparison of the two projections.

Technique 4 — Modeling of Interactive Processes. Especially in making influenced predictions and forecasts, the analyst becomes concerned with the decisive interactions among various systems and forces in a strategic situation. The basic concept is that the entities in situations such as crises will interact; they will respond to each others' actions. Thomas Belden has modeled this phenomenon as a *decision stairway* expressed in terms of time (horizontal axis) and probability (vertical axis). To take a

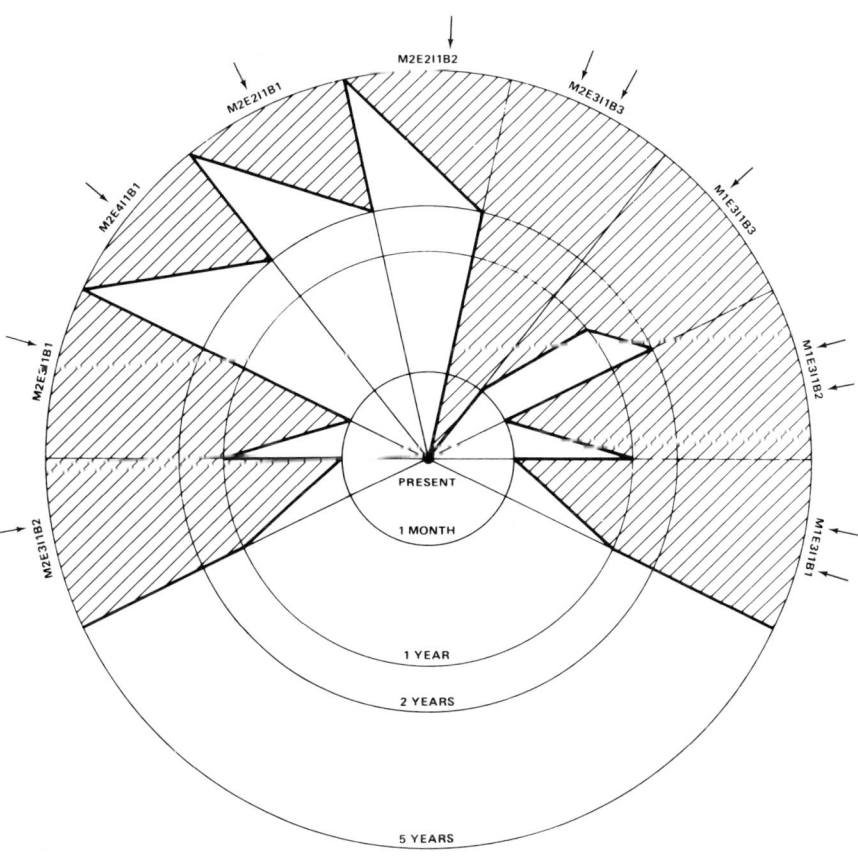

Figure 18. Example Probable Futures Map.

classic case, let us suppose that one military force is intent on launching a surprise attack on another. Assume that the attacker must conduct reconnaissance and perform various logistical and force mobilizing preparations before he is ready to attack. One can think of these acts as forming a sequence leading up a staircase toward increasing readiness to attack and, finally, the decision to do so. If at some point the potential victim detects such preparations and reacts on that basis (for example, signals awareness), the attacker may be deterred or halted and proceed back down the stairway and away from readiness to attack and the actual decision to do so. In making projections, the strategic analyst must be attentive to the critical impact of such interactive processes on likely outcomes.

Technique 5 — Rationale Recording. What I have been describing are some recommended forms and techniques of strategic projections. Obviously they are designed to mitigate problems in human inferential strategies and to foster realism. I have developed a general framework as follows:

- Projections ultimately channel into the basic information categories.
- The state of social systems and other entities may be considered in terms of the status of certain of their sectors. Various states of these sectors are more or less likely in various combinations. These combinations essentially amount to sets of conditions.
- The likelihood of given sets of conditions at various future points should be estimated with recourse to several basic principles affecting societal dynamics.
- Possible situations of strategic importance should be "overlaid" onto the mappings of possible conditions.
- The mappings should be used as an aid in discovering new situations of concern.
- The mappings should also be used as an aid in postulating the most likely unimpeded conditions out to various future points and the implications for policy.

But the question remains, What about the specific rationales behind the projections? Exactly what rationales can analysts use in systematically worrying projection from the viewpoint of the basic principles of societal dynamics? How can they resolve the inevitable tensions among the competing rationales?

There are a number of approaches, ranging from qualitative to quantitative, and sometimes involving a blend of the two. Models of societal change and international conflict may be constructed on the basis of paradigms of psychology and sociology. As another approach, argument by historical analogy may be made; for example, detailed models of past crises and strategic surprise might be constructed as a basis for recognizing future threats. In still another approach, Bayesian, Markovian and other methodologies based on statistical techniques involving probability could be used.

There is, however, little doubt that we are in the infancy of the development of strategic projection techniques, a point directly or indirectly implied at every point in the present research. As just one example, the critical need in strategic analysis to understand better the psychophysiological constraints on foreign leadership behavior, especially in a predictive sense, is largely unmet. There is no question that this is a huge problem with substantial political overtones, but progress has been far too limited. In a report prepared for the Defense Advanced Research Projects Agency, Gerald W. Hopple states:

> . . . the elucidation, measurement, and analysis of an array of individual- and elite-level factors should enhance our capabilities for . . . *estimating* and *predicting* the intentions, preferences, and probable behaviors of other actors in the international arena.

But Hopple must then note that although there has been considerable research in the various relevant fields, the following is reality:

> As is customary in social scientific inquiry, however, the existing work is disparate, uneven, and *ad hoc* in nature. Few efforts have been undertaken to map out the terrain in more than a cursory fashion.

It is not my purpose here to survey and compare various specific projection techniques currently available; there is a considerable literature on these techniques. From the present perspective, what I *do* say about them is two things: first, they should fit usefully into the basic framework of strategic projections; and second, they should be viewed as heuristic. For we may be certain that there is now nothing like a fully satisfactory projection technique for strategic analysis. Improved techniques must be developed. As is being argued in these pages, one of the keys to such development lies in extrasomatic memories and the learning processes they facilitate.

Regardless, we are concerned with the capturing of whatever rationale is used by strategic analysts making projections, no matter how qualitative or quantitative, rudimentary or sophisticated; for rationale must be recorded for purposes of measuring signification (and ultimately for operations research in *post mortems* to identify technological and methodological needs).

We may consider rationale recording at two levels: the level of conditions and the level of threats. The formats of the rationale records should not differ significantly between the two. At the first level the analyst is recording the rationale behind the conditions he has projected will hold in a given country or sphere of interest at a given time. Table 9 is a matrix relating rationale to the projected conditions and the principles of change. The matrix, which has not been completed, is intended as a simplified example. The statements of rationale can be as technical and detailed as warranted and the analyst is capable of producing. In a computer-based memory system with graphics displays, obviously there must be abbreviated rationale statements for quick display and review, with access to more detailed back up statements, some of which may not need to be stored in the computer (for example, "fixed" rationales such as those in Bayesian approaches).

Useful supporting statements concerning the basic assumptions behind both levels of rationale recording — conditions and threats — should be employed. Very simple formats for such structures are provided in Table 10, Panels A and B. (While the formats are simple, the accompanying statements of rationale can be detailed and substantial.) One possible technique is a vector approach (see Panel B) in which the analyst, by whatever rationale and methods are available to him, develops vectors whose length is a measure of his judgment of the relative impact of the various (sometimes competing) principles of change. The analyst might "resolve" the vectors and formulate a projection.

Various other structures of rationale are applicable. One further example concerns rules for making probability judgments in strategic warning. Thomas Belden has described some rules as follows:

a. *The more precise the prediction, the lower the probability.* If one says A will attack B on 16 September, that statement will have a lower probability than, A will attack B in September.

b. *The greater the number of information elements within the probability statement, the lower the probability.* For example, three of A's divisions at X will attack two of B's divisions at Y on 16 September will have a lower probability of being correct than A will attack B on 16 September.

Sector	Internal Self-Consistency	Similarities	Cause/Effect	Goal-Seeking	Holistic Trending	Continuity
M2	Cannot expect large-scale military expenditures over one month given state of economy and political machinery for doing so.		Not enough time to change to M1 or M3 states. Could not purchase/build sufficient military capability.	Goal is to occupy/dominate entire peninsula. Therefore, no slackening of military capabilities envisioned.		Has not in past demonstrated capability to make that rapid a jump. Not typical.
E3		Is similar to a number of other communist/noncommunist.	Could be nothing in single month likely to cause.	Have new 7-year plan, but do not envision much change over one month.		Rate of change not to be very discontinuous in one month.
B3			Nothing being done by U.S. or ROK to cause change in B3.	Given strong goal of reunification, the failure to launch military attack betokens B3.	Visits of PRC high-ranking delegates for most of month. Expect stable posture.	
I1				Goal remains status quo.		

Table 9. Example Rationale Matrix (One Month — M2-E3-I1-B3).

c. *The overall probability of the statement cannot be greater than the probability of any one element.* Using the first example, in b. above, if there is only a 30 percent probability that B has two divisions at Y, then the overall probability cannot exceed 30 percent.

d. *In general, the greater the time span of prediction, the lower the probability of its occurrence.* This is so because many events can intervene to change the situation. This requires that all probability statements have the date the prediction was made.

Analysis Routine. We should now consider how the methods and techniques of making strategic projections correspond to the steps and procedures in the model of strategic analysis described in Chapter 2. Upon review it is evident that use of the forms and associated procedures for projections will lead the analyst through the full steps, procedures and subprocedures identified in the model. By focusing on the basic information categories, the analyst can translate threat models from the preceding stage into projection formats. Using the Specificity Rating Scale, the analyst can develop and compare the specificities among various threat estimates. There are also approaches by which the analyst can develop rationale for projections. In short, sector comparison, probable futures mapping, modeling of interaction processes and the other approaches represent an integrated analytic routine which assures that the analyst will think very systematically and with considerable mental energies about the prospects for various strategic situations. For example, he will question hypotheses, generate new ones, seek disconfirming evidence and, in general, undertake directed devil's advocacy. This routine constitutes a defense against the epistemological and cognitive problems in strategic analysis. The backlog and relationality measures are used to monitor the degree to which the analyst pursues the methods and techniques and follows the steps and procedures in the projection stage.

More generally, we may also observe that the system of analysis from monitoring to projection is a coherent analytic routine which culminates in a set of judgments concerning which situations are most likely and least likely at any given time. We can think of the analyst as always confronting a bounded set of possibilities and attempting always to have identified the more probable of these and related them to potential situations of strategic importance.

Signification. In the discussion of the threat recognition stage, it was pointed out that signification is indicated when there are changes to and in the threat models, for example, the association of new data with

Table 10. Example Rationale Formats.

```
DOMINANT SECTOR:
    ECONOMIC
ORDER RATING OF REMAINING SECTORS:
    B,M,I
DECISIONMAKER PREDISPOSITION:
    CAUTION, OPPORTUNISM
DOMINANT CHANGE PRINCIPLE:
    CAUSE AND EFFECT
ORDER RATING OF REMAINDER:
    CONTINUITY
    INTERNAL SELF-CONSISTENCY
    SIMILARITY
    HOLISTIC TRENDING
    GOAL SEEKING
```

PANEL A

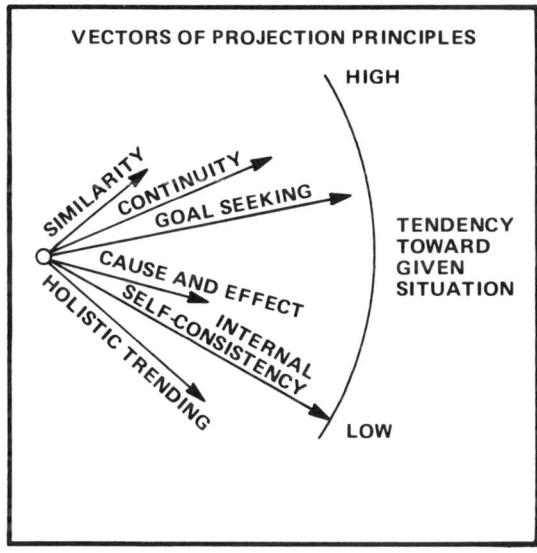

PANEL B

nodes in the models such as the critical event filters. It was also indicated that the full meaning assigned to new data is bound up in the assumptions underlying the models. A distinction was drawn between high and low context analysis and their interactive relationship in the process of strategic analysis. It was also seen that in the projection stage, changes in meaning that indicate signification can be recorded at a number of points, from the specificity ratings through the sector comparisons, probable futures maps, decision stairways, rationale matrices, rationale vectors and so on. Both changes *to* and *in* these analytic structures may be recorded. Further, to the extent that the assumptions underlying the threat modeling in the threat recognition stage are developed in the projection stage, signification involving development of, and changes to, such assumptions cross relates to the former stage. In short, there is obviously a considerable structure of signification inherent within the full techniques and methods of analysis from monitoring to projection.

The techniques and methods used in all three analysis stages, and especially in the projection stage, culminate in diagrammatic and alphanumeric formats which represent a sort of scoreboard for signification. The outcome is a series of machine-driven display formats used to present recorded signification values at various levels. The potential is there to examine the signification values in a small portion of a large analysis or to display in sequence a detailed and extensive summary of signification in a large-scale piece of strategic analysis developed over a lengthy period. Here the objective might be to trace the evolution of analytic perspectives, showing the impact of input data on the rationale and assumptions of the analysts; and *vice versa*, analysis being a tug-of-war in which the analyst struggles to fit the data to his analytic structure with mixed success, the failures causing him to review and possibly recast portions of the analytic structure.

The actual mechanics of monitoring signification and the other variables are comparatively straightforward. I will outline an approach involving computer-based operations, one that would be typical for applications of cognitive technology. I will consider backlog and relationality as well as signification.

First, we may examine *backlog*. Backlog would be monitored for each analytical stage and each step and procedure within stages. To measure backlog, the computer-based system must track and monitor a number of elements, including the following:

Stage	Function
Monitoring	Identification, date and time of arrival of data items. Times at which analysts reference data items. Times at which indicators are generated or identified and declared by analysts.
Threat Recognition	Times at which indicators are related to applicable threat scenarios (I.e., individual nodes and elements).
Projection	Times at which the output data from the threat recognition stage are reviewed against the projection structures. For example, the times at which the results of the threat recognition analysis are reviewed for their impact on sectors, the various projection maps, etc. The times at which these activities are completed.

Relationality also should be measurable at each point within each major stage of the analysis process. As discussed in Chapter 2, the model of strategic analysis encompasses a number of steps and procedures in each of the major stages. A series of methods and techniques of analysis have been defined in the present chapter whose enactment results in the analysts. carrying out these steps and procedures. The basic design approach for developing relationality values is to make sure there are recordable man-machine interactions at critical points of activity in such analytic sequences. The analyst would log on and off the system and perform certain analytic steps and procedures manifested by the calling up of displays and other activities. By recording such man-machine activities and timing them, the relationality measure would be derived.

Stage	Monitored Activity
Monitoring	Call-up of newly received input data for review.
	Categorization of input data.
	Call-up of predetermined indicator lists.
	Identification and display of correlations.
Threat Recognition	Call-up and display of preestablished threat models (PAMNACS, DENs, etc.).
	Call-up of displays comparing threat models to which data has correlated.
Projection	Call-up of threat predictive formats.
	Call-up of specificity rating scale.
	Call-up of sector lists.
	Call-up of probable futures maps.

Signification should also be measurable at each important point of analysis. I discussed earlier the strategy for measuring signification. The basic mechanism is to monitor and record changes made to the models and other structures as analysts assign meaning on the basis of incoming data. Examples of the items to be monitored are the following:

Stage	Monitored Activity
Monitoring	Changes in the activity levels assigned to indicators. Changes in the priorities or weightings given indicators. Indicators not called up for review against given input data.
Threat Recognition	Changes to the nodes or other elements in the structure and forms of analysis. Association of given indicators with specific nodes of threat models.
Projection	Changes in the specificity values assigned to given projections. Changes in the predictive formats. Changes in the number of sectors. Changes in possible future maps. Changes in assumptions matrices.

Since we would determine the relative times spent on various activities, and the nature of those activities, it would generally be possible to develop a considerable amount of insight, based on monitoring a large number of values. A few of these are:

- The time since each scenario, model or other piece of analytic structure has been modified.
- Time spent on various activities.
- Changes in complexity and dimensionality of models.
- Times between management reviews of analytic status in various problem areas.

Of course, there are deeper aspects of signification, and we have explored both the measures and the processes of strategic art sufficiently to examine some of them.

First, it is important to review the general concept of strategic analysis. I have been viewing strategic analysis in part as a process by which the imagining of the future is organized to focus on estimating ranges of political, economic, social and other conditions, and from there to imagining plausible situations of strategic concern within these constraints. In some fashion, comparative probabilities of occurrence must also be established.

I discussed earlier the need for "disciplined" imaginative analysis. The use of the six principles of forecasting, and the search for credible ranges of conditions — really, the search for the most likely possibilities in a world of many possibilities — constitute the essence of the disciplined imaginative approach in strategic analysis. (Opinions will, of course, differ as to the appropriate degree of constraint to place on the imagining of future states.)

Within this model of strategic analysis we find the analyst employing intuitive inferential strategies — judgmental heuristics and knowledge structures. We find him running certain risks in such inferential strategies in estimating causality, making predictions, determining covariance and in other analytic tasks. Since we also intend to employ the machine in support of the analyst, problems such as those bound up with the availability and representativeness heuristics can apply both to the human analyst and to the machine and its memories and algorithms, since the latter are developed and programmed by humans. The machine, used to extend the capabilities of the natural human inferential strategies and to offset some of the problems of unaided human inferencing, must itself be increasingly well-understood and refined through *post mortems*.

I discussed above some contemporary cognitive research that is exploring problems in human inference. For example, researchers have explored such problems as the fact that often predictions made by unaided human inferencing fail to take account of regressive factors. Similarly, problems appear to arise in the determination of causality because the vividness factor sometimes disastrously influences what information is most readily available for use in judgments. Ideally the machine-based *post mortem* memory which holds signification events must afford us some chance of reviewing problems in availability both for humans and machines. For example, we must hope someday to be afforded clues as to whether in attacking strategic problems human analysts take precautions to offset the possibility that the information immediately available from their own memories is inappropriate. We must determine whether analysts systematically augment that information by recourse to other information less readily or immediately available, including information within machines. Here we should note the importance of the backlog variable, for it might allow us to determine whether (1) inadequate information is available in the machine even though useful information has been received at an analysis center, or (2) appropriate information is simply not available at any point. More basically, one of the great challenges in developing *post mortems* will be to determine the extent to which the machine memory and the procedures of analysis work against the natural traps in unaided human inferencing.

An especially important problem to seek to understand better through *post mortem* memories of signification is the limits on our ability to imagine future states. A basic question is, How well can we imagine what we have not directly experienced? Here one subtle issue is the organization and programming of machine memories. Machine memories must be developed to foster the interpretive impulse, particularly the imaginative thrust of strategic analysis. If memories are organized efficiently and the processing of their data for recall developed to facilitate rapid, effective man-machine interactions which allow and encourage analysts to examine many different combinations and possibilities of processes, dynamics, conditions, etc., then we might achieve a cognitive technology which fosters the interpretive thrust in strategic analysis. We must strive to develop *post mortems* which give us visibility into these kinds of subtle problems and their solutions.

Examples of the types of questions of interest include: (1) How many man-hours were spent at various stages and steps of strategic analysis?; (2) What were the nature and content of the memories available to the analyst?; (3) Typically, what portions of the memories were most used and which least used?; (4) What were the reasons for such usage of memorialized data (e.g., is a type of "vividness" factor at work in the display formats or machine programming?); (5) What kinds of refinements were made, and how often, to model structures?; (6) How often and in what ways did the machine support help the analysts to use "pallid" data as well as the more "vivid" data available through their own memories?; (7) How many hypotheses were formulated?; (8) How explicitly were assumptions described?; and so on. Although these may not be the subtle questions we will be asking years from now, nevertheless they are important questions. And there is no doubt that with even a primitive system of analysis such as that outlined in this chapter, together with programming that is now quite feasible, they can be readily addressed through the cognitive technology possible today.

Even so, we will not easily learn how to make analytic routines, modeling procedures and display format designs play together to help us tabulate, interpret and understand the dynamics of strategic analysis; measure the improvement in the generation of hypotheses and the assumption of devil's advocacy roles made possible by the speed of operation of the machines; and assess the potential through recourse to machine memory of creating vital analytic context. But we must try very hard to do so, for it seems more than reasonable to judge that with machine-supported strategic analysis, and requisite man-hours and intensity of effort, we may do a *remarkably* better job of imagining realistic probabilities of strategic change than we might have believed we could a decade or two ago. In short, there can be no greater imperative in the development of cognitive technology in support of strategic analysis than to proceed to invent ways of making better memory systems.

Cognitive Technology Implications. In earlier remarks on the implications of cognitive technology, I discussed the general need for requisite speed of operation. In the present discussion another related and extremely important requirement becomes apparent: computer displays and display techniques.

Strategic analysis must include a substantial *visual* dimension beyond oral and written traditions and records. This visual dimension must be largely electronic in nature. That we need to operate substantially in a visual mode in strategic analysis arises from several considerations I have argued above and which may be summarized as the need for speed of operation both to aid analysis and to aid measurement of analysis.

Implicit is the theme that computer display technology should be harnessed to allow analysts to operate with extrasomatic support systems designed to complement both the cognitive processes of analysis and the measurement of analysis. Analysts must be allowed to create extrasomatic memories in the form of certain models or schemata, and to manipulate them readily through interaction with the machine, for purposes of changing the structures, recording analysis processes and results for measurement, and recalling the analytic memory. The structure for analytic memory I have described above — PAMNACS, DENs, PFMs, rationale matrices, etc. — is coherent and comprehensive; but without machine support it could not be easily recalled and manipulated by the human analyst. Given the machine, however, the structure becomes readily usable for analysis and its measurement.

To this point I have examined various elements in a computer-based system of strategic analysis: speed of operation, measures derivation, graphics, etc. It is now useful to step back and view briefly an overall architecture for such a system. Figure 19 depicts the basic shape a system might take. I will comment on some of the elements in Figure 19.

Let us begin with the *Analysis System*, the structure described earlier in the present chapter. First, consider computer support in the *monitoring stage*. I have noted that the analyst may be supported by various computer-based qualitative and quantitative techniques for monitoring indicators. Preferably the analyst will be able to display lists of key indicators. In addition, quantitative analytic techniques which involve assigning priorities and weightings to various indicators and combinations of indicators may also be available. Whatever the status of such support, the strategic analysis system should be designed to assist the analyst in whatever functions are entailed, such as (1) review incoming data applicable to his assigned activity or region; (2) extract the pertinent information; (3) correlate the information against indicator lists; and (4) display results. Depending on the quantitative techniques available to the analyst, the result generation may involve a statistical presentation.

With respect to the review of received data, the analyst, having stored previous data in the system, can readily extract and display updated indicator lists. The system should furnish assistance by providing current activities, indications or data points that must be verified, checked or otherwise considered. This function gives rise to the important concept, already noted above, that the system include the capability through its graphics to prompt and cue the analyst by means of variations in color and other techniques such as flashing and underscoring. The prompting should involve such functions as helping the analyst determine what steps and procedures in the analytic routine might be advisable as a result of previous correlations.

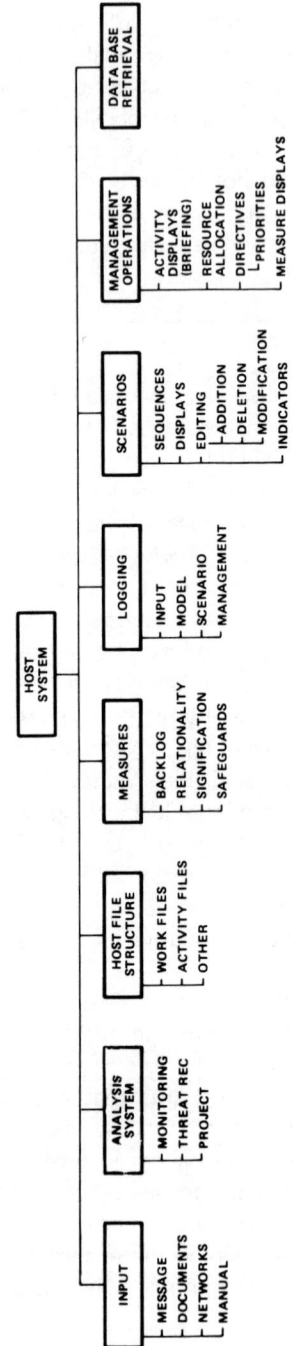

Figure 19. Design Elements.

Now let us consider the *threat recognition stage*. Considerable detail was provided above on this stage. The analyst seeks to account for indicator activity from the monitoring stage by reviewing the activity in the context of an interrelated set of artificial schemata or models which structure the primary situations believed pertinent to the strategic problems in question. These models range from large-scale depictions of possible courses of action through increasingly detailed depictions of the specific activities and events that would characterize the unfolding of the major actions. The software should support analysts in a number of ways, including the following:

- The threats modeled in scenario form may be stored and called up.
- In the display of models there are indications of the status of models, for example, indications of past instances when certain indicators were linked to different nodes. Also included are prompting mechanisms which help increase the comprehensiveness of analysis, for example, prompting which aids the analyst to review the different models systematically against the indicators moved forward from the monitoring stage.
- The analyst can change or modify existing threat models on the basis of received data; he has the ability to generate completely new models through a configurational language.
- There are graphic indications of the ramifications of signification. Shortly I will discuss the fact that when the analyst changes given elements or nodes in a model, often the change ramifies through the logic structure of other related models and carries the implication that changes must be made there as well in order to depict the full meaning that has resulted (and which ultimately brings changes in signification values). The system should at least prompt the analyst to proceed through a series of review steps in which he examines other structures that might be affected by a change in judgment at a given point.

The role of the strategic analysis system in the *projection stage* is essentially the same as in the threat recognition stage. The difference, of course, is that the computer-based support is now directed to the projection maps, matrices and other structures used in projection. Capabilities for storage, modification, display and recording would be provided for these elements of projection:

- Statements of predictions and forecasts in the format of the basic information categories.
- Specificity rating scales, using color and other techniques to emphasize levels.

- All appropriate computer operations on sectors and their comparisons.
- Single sector and probable futures maps; and rationale matrices.

It should be possible to review all of these outputs as to past versus present status. Analysts should have access to such summaries through the system of displays.

I must also emphasize that the system should incorporate *scenario language* for use by analysts. The term, scenario language, refers to all elements within the total structure of models, projection maps, assumptions matrices and other forms within the analytic context. The scenario language which would be provided for the analyst may be functionally described as follows:

- Block and node oriented.
- Parallel activities/sequential actions.
- Scenario building blocks provided.
- Scenario editing available — addition, deletion and modification provided to analysts.
- Scenario display via graphics, with color and dynamic prompting.

Finally, I note that a rather straightforward set of requirements should exist in such a system for file architecture — analyst work files, activity files, etc. — and for manual and automatic data entry.

One classic illustration of the importance of computer-based support is the case in which the strategic analyst must review and respond to a probing, broad estimate of likely developments in his sphere of interest. Intelligence estimates made by the government of possible developments in various world regions are examples of such studies. These substantive, far-reaching estimates, made by various agencies and offices, are prepared periodically and should be utilized by government strategic analysts in developing high context analytic structures of models, assumptions, etc.

Let us assume that an analyst reviews a detailed analysis of the political, economic, military and other constraints on both North and South Korea with respect to policy over the near term. The estimate probes the interrelationships among economic problems, political problems (domestic and international), military force shortfalls and numerous other aspects, particularly those influencing the policy postures taken by the leadership in North Korea. Let us also assume that regardless of the extent to which the given analyst might subscribe to the assumptions and judg-

THE ART OF STRATEGIC ANALYSIS   139

ments in the report, it is generally a very comprehensive and useful document.

This estimative report (and others like it) would ordinarily impact strongly on the models in the threat recognition stage and on the structures and assumptions in the projection stage. For example, assume that the analyst in question adds a new sector to the list of sectors he earlier developed for North Korea. Assume that prior to this the analyst had established these sectors: E (economic status); B (perceptions of North Korean decision makers); and M (military force status vis-a-vis South Korea). Assume that the new sector is designated, I, the degree of internal political control exercised by the current North Korean regime.

An important point is that the addition of this new sector creates ramifications throughout the remaining structures of the projection stage. For example, a new single sector map must now be created to accommodate the new sector, and this will create a general shift in perspectives. Further, the probable futures maps, which encompass all sectors, must be changed. In turn, the rationale matrices will change, as will the assumptions, and so on.

We now arrive at an extremely important aspect of strategic analysis: ramification in meaning. What I have just described is a process whereby an analyst's decision about signification — about a change in meaning — occurs (as it must in the sequential world) at a *single point* in the analytic system. It is focused on some element in the high context system. These points now become critical:

1. Whether or not it is noticed and recorded, it may be that the change at the single point in the structure will *to the degree that the whole structure is coherent* result in far-reaching ramifications in meaning. In other words, the potential now exists for complex new meaning. In effect, then, we can think of an extrasomatic memory as incorporating a coherent set of perspectives, with this set changing to some extent as the result of a human analyst assigning new meaning somewhere in the set.

2. There is a positive relationship between the principle of coherence, as applied to the system and structures of analysis, and the dynamics of the ramification of meaning.

3. Among the specific cognitive problems of the unaided human mind are its limitations in "seeing" the extent of ramifications in meaning, particularly in timely fashion.

4. The real issue becomes that of aiding the mind by using a computer-based system programmed to (a) indicate with color-based or other prompting cues when ramifications are probably consequent, and their likely pattern, so that the analyst may realize it and explore them, and

(b) as feasible, contain algorithms which develop, record and display the ramifications. Of course, in many cases it will be the analyst who, once prompted, is obliged to confront and resolve as best he can the ramifications (or implications) of individual acts of signification.

5. These factors put into better perspective the requirement for cognitive technology to aid the human mind in its attempt to perform complex and extremely difficult strategic analysis. We now have additional context for the earlier discussions of signification.

Transition. Although the system of strategic art I have described above is primitive indeed, it is at least a beginning. There is really nothing else I want to say about it in this context. Moreover, it suffices to form a backdrop for the next subject I will discuss: the strategic analyst himself.

5
The Strategic Analyst in the Future: Speculations

If players are progressively better, a prime reason is that they had Tilden to build on. It was he who first made a science of tennis, mastered every stroke, conceived every strategem. The greatest talent and genius resided in one . . . . For every stroke he had at least three different varieties, for every attack a defense.

He could rattle on for as long as anyone would care to listen about how much spin to use to answer each particular shot, how to achieve 'the middle road of spin plus speed,' and where exactly to place the racket upon the ball (southeast corner for a forehand, southwest for a backhand). Ideally, he hit a ball waist high, just as it began its descent, and, like snowflakes, no Tilden shots were ever quite the same. 'The great majority of players are not students of the game,' he wrote.

—Frank Deford, *Big Bill Tilden*

What Bird did was to synthesize the discoveries of all the searchers for new sounds and, like the true innovator in an art form, show others how to put the off-beat, the far-out and the weird together so that they became aesthetically pleasing.

—Arnold Shaw, *Charlie Parker: Segue to Today*

Modern strategic analysis is sure to have its innovators, its geniuses of technique and method. Future virtuosi will invent more elegant and economical techniques of analysis. At ease with the machine, they will create better ways of programming and using the machine. They will invent training systems which foster cognitive fitness. Above all, they will improve the measurement of real effectiveness, for in challenging reality they must learn ever more about their inferencing abilities and hence must diligently seek to design better *post mortem* memories yielding more insights into these processes. Such understanding will obviously also be bound up with the motivation, the excitement, of the quest, for it will be the foundation for judging achievement. And it will define the requirements for mastery.

It is my strong belief that the genre of strategic analysis will grow dramatically in popularity and appeal, attracting to its pursuit some very bright and talented people, and a few geniuses. There will no doubt emerge different styles, movements and periods. And certainly there will be triumphs as well as failures.

The great successes in strategic analysis will be linked closely to their times, places and conditions. Superb analysts of one generation, like great artists and athletes, will have to be judged partially in terms of their particular history. What will be common to all periods — what will create the motivation and bring forth the genius and energies of the innovators — will be the enormous challenge, the great mental expeditions and explorations, involved in seeking to project reality.

I have suggested that strategic analysis is basically an art, embodying an element of the performing arts. I have stressed that the strategic imagination becomes the major hope of successful strategic analysis; that we have no alternative — certainly not science — other than to turn the techniques and methods of art, intensified, expanded and enlarged by cognitive technology, to envisioning the future, stirred by the imperative of looking ahead in the name of survival. Much ultimately depends on the strategic analysts themselves.

In my view, strategic art will become more refined and sophisticated largely through the discoveries and innovations of the *practicing* analysts. The nature of strategic analysis — its focus on an everchanging, always novel reality — is to demand from its practitioners opportunistic and ingenious invention of technique. It poses the challenge to develop innovation in mental skills, skills in performing extraordinarily difficult inferential tasks. Indeed, it is among the greatest challenges to the human mind. For it challenges us not only to surpass our immediate, local environments; and not only to explore well beyond the immediate meanings of new information; but it tasks us to reconnoiter the unhappened, to imagine the future accurately.

THE STRATEGIC ANALYST IN THE FUTURE: SPECULATIONS

Herewith, then, I offer a few speculations (including those above) about the strategic analyst of the future. My speculations are presented in no special order and, as will be obvious, are far from evenly developed. But I believe they touch on a number of basic issues. If they cause some readers to ponder such issues further, and to discover more important ones, I will have achieved my purpose.

In a period that has seen modernism and postmodernism in the arts, strategic analysis, supported by cognitive technology, may well become a new genre. It will have its own aesthetic. The aesthetic of strategic analysis links the artistic impulse behind writing, painting, music, moviemaking and other art forms — the imaginative creation of models — with the imperative to see ahead with literal realism.

But there will be enormous challenges. Indeed, in the view of some there is now an unprecedented epistemological crisis which throws into doubt our interpretive ability, our success in making sense of events, our ability to project outcomes. Two of the sources of the crisis most frequently cited are, first, certain discoveries in twentieth century physics; and, second, the impact on humans of advances in communications and electronics technology.

Within the present scope, I can do no more than briefly suggest some of the epistemological impact of twentieth century physics. With considerable arbitrariness, we can think of several historical stages, beginning before the century and dating back to Newton:

Stage 1: The physics of Newton produces a new model of reality, a new epistemology. A scientific view of the real world emerges in which there is a coherent system of predictive principles. In the eighteenth century, Laplace claims that now the universe can be described with confidence; that it is predictable with certainty. Concepts of man in his perceptual and cognitive relationship to reality tend to hold that there are "facts," and that these are epistemologically reliable.

Stage 2: In our century Einstein throws into question Newton's gravitational theory and its perspective on temporal and spatial constraints, its reliance on laws of classical mechanics. If "facts" were previously considered reliable, Einstein's counterintuitive revolution in physics now creates uncertainties. In addition, the concept of the elementary particle is developed. Quantum physics, which revises the principles of classical mechanics, alters the older notions of a universe ordered in a predictable system of cause and effect. Heisenberg's concepts of uncertainty further undermine the perspective of the older physics, creating a concept of a microworld of particles characterized by probability and randomness. Niels Bohr's concepts of "complementarity" justify an accommodation with ambiguity. Bohr argues that opposing theories of the nature of matter are valid and coexist in a complementary way.

Stage 3: The realization grows that there is a world underneath obvious causality which is not graspable through current means of information; that earlier deterministic views have become suspect at the very least; and that the appropriate attitude seems to be a possibility oriented attitude toward reality. As one observer has written:

> The behavior of the particle is uncertain and therefore the behavior of the atom is an uncertainty. The behavior of the atom can be predicted to a degree of probability. The behavior of an aggregate of atoms is therefore only a probability, and not a certainty. And it is no use saying that the degree of uncertainty is too small to affect events on the ordinary scale, for the notion of determinism is similarly based on the fundamental determinism of the individual molecules, multiplied many times to become the world of nature.

Hence in the final analysis, the new theories have a broad impact on man's confidence.

> (George Smiley) hated the Press as he hated advertising and television, he hated mass media, the relentless persuasion of the twentieth century. Everything he admired or loved had been the product of intense individualism.
>
> —John Le Carre'
> *Call for the Dead*

Perhaps of even greater significance to strategic analysis is the impact of the modern communications media. The following is a typical view:

The new communication technologies make the formulations of any encompassing authoritative visions increasingly more difficult, since they produce an information overload which gives such diverse and disparate views of reality that no single interpretive frame can contain them all and still present a coherent vision of experience. The information revolution also expands the range of the probable to the extent that it blurs the boundaries of fact and fiction . . .

The strategic analyst obviously is victimized by the information deluge. Modern communications media have created a huge collagé enlarged daily but perversely discontinuous. It is a collagé of facts, images, events and large and small "stories." Beyond doubt it is an information overload of diverse and disparate views of reality.

The human neurological and psychological responses to this information deluge, particularly in professional observers and interpreters such as journalists, strategic analysts and writers, is sometimes described as a kind of pathology. The specific problem becomes the inability to connect things, to impose order, to narrate process. On emotional and motivational levels, the result can be a variety of inappropriate adjustments: indifference; escape through channeling the interpretive impulse to the trivial; demoralization; and paranoia: in sum, the symptoms of a crisis of interpretive confidence.

Here I must ask for the reader's indulgence, for the impact of media on the interpretive impulse is so important to our discussion that we must briefly explore it. Probably the most dramatic (if among the most extreme) symptoms of the crisis are some recent developments in American literature, developments whose significance is not unrelated to the strategic analyst. In the last two decades, we encounter some novelists who blur the distinction between "fact" and "fiction." There has been considerable change in narrative forms: nonfiction novels, transfiction, surfiction and other new forms designed to mix the real and the fictional. The chief reason, as one literary observer suggests, is the view that "the official level of reality is so weird that what passes for realistic fiction is totally anachronistic, like an old road map" (essentially a variant of the old adage, "truth is stranger than fiction"). Thus some novelists view their role differently now: they want to be "directly involved" with reality. "Purely imaginative" works can no longer compete with reality. In this sense, then, the writer is becoming more like both the journalist and the strategic analyst.

A dominant source of these new perspectives is taken to be the modern communications media. The media have amplified the fictive or fantastic quality of reality. Because the major media are driven by economic considerations — are, in fact, in show business — the news tends greatly to emphasize "stories" such as Vietnam, the America of 1968 (assassinations, urban ghetto fires, campus wars, etc.), Watergate, and Iran of 1979 and 1980. There is judged to be a circus quality about the major media reporting. Through the mass media we are used to seeing images — usually disconnected and random — of spectacular events and people caught up in the centers of power, which in turn are the standard stages often of what strike many as rather fantastic and bizarre situations and behavior. Because of time, space and economic constraints on reporting and analysis, the output of media remains a collagé whose meaning is difficult to determine. Worse, the long-term effect is to condition one to the present and to the superficial explanation.

Among some contemporary writers, the reflective tendency has been to *reduce* the interpretation of reality. Hence even though they are becoming more "journalistic," these writers are becoming far less interpretive. Writers working in the new forms tend to approach the world "as it is" without imposing a framework of meaning. The French writer, Robbe-Grillet, has said that reality as experienced by us is "neither significant nor absurd. It *is*, simply." Some modern literary heroes tend to end up either having lead ambitious expeditions to explore reality which fail to yield much meaning, or simply having decided not to set off on such journeys at all. Writers frequently cited as working in the new forms include Thomas Pynchon, Donald Barthelme, John Barth and Steven Katz.

Such writers depart the tradition of classical novelistic models written on the assumption that it is possible to have an integrated view of existing realities. Instead, they attempt a neutral registration of experience, rejecting the concept of art as the creation of order from chaos and the writer as an interpreter. Such writers do not approach experience in search of cause and effect; they do not organize experience linearly, with logical development and the striving for resolution. Techniques of collagé are more likely, since the writers are opposed to more traditional novelistic forms which seek to perceive and interpret a whole from the parts of reality experienced. In a recent study, Mas'ud Zavarzadeh has interpreted Thomas Pynchon's novel, *V.*, as an expression of the mood, the feeling, of such writers in their pessimism about meaning:

Mistrust and even fear of totalization not only inform the

attitudes of innovative American narratists today but also serve as their immediate subject matter. Such dread is the basic thematic motif of Thomas Pynchon's *V.*, in which the protagonist, Herbert Stencil, ostensibly seeks a literal synthesis of the numerous confusing manifestations of V., but actually fears any solution to 'the V. jigsaw.'

(Stencil wishes to avoid) the false assurance of having obtained the harmonizing principle behind the manifold reality. Stencil's fear of piecing together the hints, clues, and signs he has come across during his search — caused by his suspicion that these are all planted in his way by some conspiracy to mislead him and prove that his integration of V. is no more than merely a scholarly quest for a synthetic wholeness — is opposed by the equal weight of his anguish in not being able to locate V. and separate her from various V.-like appearances. His conflicting feelings . . . are indications that while he has a strong yearning for a unitying belief, the norm in his life has so unmistakably become the fragmentation and chaos of experience that even the thought of reaching a cohesive vision of the wholeness beneath the scattered surfaces is not only suspect but also phony. This agonizing impasse is characteristic of contemporary man's compulsive attempt to identify a solid core of trustable reality distinguishable from fictitious appearances and his mistrust that the pattern he comes up with may be a mere projection based on no more than recurrence of an initial and a few dead objects thrown in his way by cabals. As one character in *V.* states: 'In a world such as you inhabit, Mr. Stencil, any cluster of phenomena may be a conspiracy.'

Ultimately, reality is viewed as:

. . . a form of organized chaos, resisting interpretation and inherently indeterminate to such an extent that a paralogic of conspiracy is the only kind of reasoning which can account for its strangeness.

Most importantly, a profound tension arises because one of our strongest impulses in the interpretive impulse.

Perhaps it is naive to suppose that our times are really any more bizarre than earlier times with their plagues, religious wars, ironic expeditions and rapacious colonizations. But surely twentieth century technology has made the world smaller, more dangerous. And clearly the "age of information" and "the data market" have resulted in an information overload that is peculiarly perverse. It seems beyond question that the interpretive impulse has been blunted. Confidence in the power of the imagination to simulate reality and lead to understanding has diminished.

It suits our purpose here to take account of the interpretive crisis, to recognize it as a major obstacle to be overcome. Certainly the strategic analyst does not live in a vacuum; he is affected by the general environment. Thus while I have been reviewing some developments in literature to suggest quickly the general crisis of interpretation, we should note that symptoms regularly appear in the world of the strategic analyst as well, such as:

- Skepticism (really, pessimism) at the prospects for more sophisticated and innovative analytic methods and approaches.
- Extensive emphasis on indicators and monitoring to the neglect of modeling and projection. This is sometimes referred to as a lack of proper emphasis on the estimative, long-term aspect of strategic analysis.
- Less than adequate funding for the R&D necessary for innovation.

Without the machine and the ability it gives him to elaborate and amplify his knowledge structures, to create new knowledge structures and to create a reliable extrasomatic memory, the analyst will be unable to overcome the information processing, cognitive and imaginative constraints on the range and reliability of his analysis which his unaided human inferential skills now labor under. In the lexicon of today's research, he may be defeated by inappropriate uses of the availability and representativeness heuristics; by inadequately challenged theories and beliefs; and/or by the sheer intensity of the information deluge. The skeptics will be vindicated and, worse, the interpretive crisis will deepen.

Hence the strategic analyst must come to exploit information machines, to realize the transcendent possibilities the machines open up for analysis. I believe the strategic analyst will eagerly anticipate refinements in computers and displays. He will find himself freer and better able to perform analysis as machines diminish in size without loss of capacity and power and become more portable.

In short, it is beyond question that both the aesthetic and the future of strategic analysis are inextricably bound up with cognitive technology and the machine.

Let us consider further the aesthetic of strategic analysis. Today we appear to be turning from some of the classic dichotomies with large epistemological and cognitive implications such as realism/idealism and fiction/nonfiction. In our own time some novelists and other artists are outspokenly seeking to capture literal history within works of the imagination, to employ novelistic and other artistic devices to objectives of high realism. With the ghost of Hemingway in the background, Mailer, Capote and others come to mind at once.

As in any other period old categories of human perspective are being challenged. There is now an awareness that we must attempt to know and learn reality using our full artistic skills. Given the considerations of the limitations of science by Hiesenberg, Einstein, Godel, and others, and the beginnings of a movement in strategic art intensified by supporting technology, perhaps a period is approaching in which strategic art, itself a kind of synthesis, may emerge as a distinct stage in our pursuit of reality.

Here is an excerpt from a conversation on a Saturday afternoon in the future between two young strategic analysts, Karl Edwards and Lorraine Miyakawa, who work for a large government department. Karl and the woman he lives with, Joan, an interior decorator, are hosting the wedding reception of one of Joan's best friends. Periodically Karl, bored with the guests and their conversations, seeks refuge by hiding somewhere in the house. He has just thought to phone Lorraine.

Back in the master bedroom, he finds himself alone. He picks up the phone from the nightstand and carries it several paces to the closet door, the long phone cord emerging from under the bed and snaking across the carpet. He opens the closet door, steps into the semidarkness and carefully places the phone down on the carpet. Then he walks to the glass door to the balcony, slides it open, steps out carefully, picks up a set-up folding chair, steps carefully backwards into the bedroom, making sure not to scrape the chair against the doorframe, and returns to the closet, putting the chair down next to the phone with the chair facing the closet door. He flicks on the overhead light in the closet and shuts the closet door nearly all the way, the phone cord preventing full closure. Then he sits down on the chair, picks up the telephone and rests it in his lap.

What the hell, he's been wanting to call her. Why not now? More privacy than he's likely to have for sometime. Today he needs some relief.

But his stomach tightens. He's probably had enough champagne to carry it off but maybe too much. Hardly knows her.

Christ, though, just *do* it!

He dials Information, listens to the pleasant recorded request to use the phone directory, and waits for an operator.

"Yes, in McLean. An L. Miyakawa." He spells it and waits.

"Thank you," he says when he has been told the number, and then he hangs up the phone.

He pictures her. Long black hair, brown eyes, soft face, a depth of reserve, an unmistakable discipline reflected in professional demeanor and businesslike conduct.

She'll cut your head off, he thinks. She's so young and she's new on the job and you work with her every day, must rely on her goodwill. You're bombed. Don't be an idiot. Probably lives with some guy.

But he dials the number anyway. There are six rings before an answer. It is her.

"Hello."

"Lorraine?"

"Yes."

"This is Karl Edwards, from work. How are you?"

"Oh, Karl! Fine. How are you?"

"I'm fine. Hey, did you go in today?"

"No. Was I supposed to?"

"No. I couldn't make it and I just wondered if somebody on our shift looked at the traffic." What idiocy.

"Sorry, no I didn't. Was I supposed to?"

"No, no. I'm hosting this dumb wedding reception here and I can't get my mind off Korea." Wow.

She says, "I almost asked you for a Korea review Friday. A couple of us think maybe you're ready to call it."

"I *am* ready to brief you people working Japan. We need to close."

"That would be great."

"How do you people see it?"

"ROK is coming apart. Our Japanese friends are trying not to face it."

"Will there be a real collapse?"

"I think so, but *When* is a good question. You tell *me*."

"Two weeks at the earliest. Six at the latest. In a few days I'll close that window further."

"How can you be so sure?"

"Am I interrupting you? Were you resting or something?"

"Something?"

"Don't embarrass me."

"I know I'm not. Besides, don't pry."

"I'm not."

Silence.

Jesus.

"Well, anyway," Karl says, "can you talk?"

"I am, aren't I?"

"OK."

"So why two to six weeks?"

"You know, it would help me if I could structure things a little with you."

"Please do."

"Can you stand some detail?"

"Sure."

"OK, start with the standard things. They all point to collapse. Stuff like more and more opposition leaders, too many to catch. Stronger and bigger demonstrations, hundreds of thousands in Seoul and other cities. Morale sinking in the military. Opposition journalists getting braver. Inflation rate going out of sight, along with unemployment. Industries almost flat. Markets shrinking. Your group knows a lot about that one."

"For sure."

"And recession. Weakening credit. Bankruptcies. And now maybe 60 percent or more of the people are real have-nots. Millions of down-and-out people wandering around, a hell of a lot of them in the cities, becoming one huge wounded but more and more dangerous beast. I haven't declined too many glasses of champagne today. Keep that in mind."

"Boldness can win the day,"

"I like you."
"Good. Why shouldn't you?"
"No reason."
Silence.
"So is *that* it? she asks. "Isn't there more?"
"Of course."
"So tell me what makes the difference this time?"
"I haven't got through all the details."
"You like to take your time, don't you?"
"There's much to be said for that."
Silence.

"I could ask a very snide question," she says. "Probably I didn't because I think I like you. But you don't seem very different." Perhaps there is a smile in her voice. He is pretty sure there is.

"It's hard to be pure about things, you know?" he says. "Things are usually all mixed-up together. Right? We have to live in our medium."

"Somehow, I sense a digression. I'd rather have more details."

"At your pleasure. You see, I *do* try. And I *know* that's important."

"How old *are* you?"

"Really, you're no fun. But all right, you can have the facts. Yesterday all regimental installations were almost empty. Armor and troop carriers building up in Seoul. Several special flights to haul out diplomatic dependents. Heavy curfews. Mass arrests. All the usual stuff. I don't need to go into it all.

"Because really, that isn't it. It's hard to describe, but it's really *acceptance*. Giving up and also taking. That point when those who used to be in charge know its over and those on the come know they can have it. Maybe both sides started to know when the losing side put Kim Dae-jung in jail. Remember?"

"Yes, I do."

"It's a very strong sense of mine that there's been a crucial relinquishing on both sides. A double momentum. The eventual winners, who thought they would be losers, have relinquished the loser's mentality. The people who are losing control have relinquished what they'd begun to suspect sometime ago was a false hope anyway, a false security. It always makes them meaner and crueler. But the handwriting really appears very early on the walls of their minds. I'm convinced of that. Lately I can just *see* those leaders in their uniforms and cornball suits sitting in the conference rooms, knowing this. And my whole conviction comes from a kind of ghostly message, a whisper, I'm hearing these days.

"As I said, I've had some champagne."

THE STRATEGIC ANALYST IN THE FUTURE: SPECULATIONS

Silence.

"Well, here's the problem," says Karl. "What I'm talking about now is entirely in my mind. I'm just not there yet, after all this time, in modeling my vision of this crisis, or others for that matter. When I first took over the Korean desk, I spent hours building this edifice everybody thinks is so great and always wants to see. It's supposed to be about as comprehensive as you can get. To reflect what I'm *envisioning*. I'm thinking of making a film of it so all the visitors don't have to bother me."

"I *thought* you might be getting tired of that."

"Do you really feel sorry for me?"

"Not really."

"Maybe I don't like you anymore."

"We'll see."

"If I promise to let you insult me at will, can we have dinner sometime?"

"I might think about it. Depends on your coherence, economy and elegance of concept."

"God, you took the course. They *all* talk like that."

"What course?"

"The one put on by MASA."

"Who?"

"Military Advanced Studies Agency."

"No, I didn't. I'd like to."

"No, you wouldn't. You'd get the lecture from that contractor guy with the red mustache and those double-breasted pinstripe suits. Dresses like a jazz musician out of the Forties. And that professor, Cumberland or some name like that, always calling everything 'awesome.' And Dr. Hourly from MASA, she sponsors the course. Wears her hair up all the time. Very businesslike. And very easy to look at."

"Add ability to concentrate," says Lorraine, "as a criterion deciding whether conceivably we might have dinner."

"Right. Let's return to details," says Karl.

"Good idea."

"So anyway, here I'm supposed to have this elegant vision of the Korean future in a series of magic diagrams, a lovely, colorful system of key variables. I can evolve through the whole thing in minutes. I've been right on the money in some events, too, though *When* and *How* are always special problems and *Why* is sometimes hard for newcomers to understand. Did you know that upstairs they call me Spiderman?"

Sho laughs.

"Yeah. I'm supposed to be like a spider moving across this large, elaborate web of ideas. It's supposed to be hypersensitive to change. Hit it with data in one area, you feel the impact at other junctions. It's hard to keep a straight face sometimes with VIPs around, hearing all that. When they're around I'll call up KORHIST, then give them a couple of DENFOCUSs and assign some Specificity Ratings and they think its a monument for the ages."

"I know. It's sad."

"Anyway, when I first tried to master this system it took me a long time. It was like trying to learn to run a marathon when you thought your body could only run five or six miles. You had to learn to accommodate a lot more complexity. There was a big learning curve. But now I'm used to it. It seems simple. And it *does* help me to be thorough. But it doesn't really match my essential visions of things. Because of the system, I'm able to build large, very convincing rationales for predictions. But the predictions are really made because of the visions and I can't trap the visions, hang them out on a line, so to speak, and contemplate them, inspect them, dry them out. So we've got to find ways for the machine to help us more. We've got to reach the point where we can get more of what we're envisioning inside the machine."

"Recover what you've imagined and also get help in the actual imagining."

"Exactly!"

"I've been thinking about this problem, too."

"Really?"

Silence.

"I've been exploring a number of new areas with my imagination lately," Karl says. "These are things I wouldn't have done if I hadn't built the system as it now stands. By doing that, I did, I admit, progress to the point where I can see new things. For example, I'm trying to imagine the experience of being in the mobs for several hours, and the sagging and renewal of spirit of the mobs. And then the elite and their daily lives these days. And young people in the ROK military. And the feelings arising from the sheer insult of the authoritarian regime, the jailings, the fear, the reducing of integrity for the preyed upon to probably fatal rebellion and no doubt a horrible, hidden death. And the hatred brought on in the attempt to manhandle and control the young with their new expectations, their new ideals and goals. Telling them they have to live their lives a certain way, locked into some gray and gloomy life of meaningless drone work, senseless work in the bureaucracy, ritualistic inheritance of already drab and dead lifestyles, telling a whole new body of people that this is how they must live."

There is a soft knock on the closet door.

"Excuse me a moment," he says to Lorraine.

The door opens to reveal the Best Man.

"Joan is looking for you," he says. "The bride and groom are leaving and also you're running out of champagne."

"Tell her I'll be there in a minute," Karl replies.

The Best Man moves away.

"Problem?" Lorraine asks.

"I'll have to talk to you again another time." he says. "I have to run."

"Sure."

"Goodbye."

"Goodbye."

I have said that the strategic analyst is obligated to imagine future conditions based on his experiential data base, against all the limitations of memorability, availability and other constraints on human cognition. He must act as a creative artist and consciously seek the greatest rigor possible. This is why the aesthetic of strategic analysis may be a blend of the aesthetic of science and the aesthetic of art. We can envision strategic analysis as having an aesthetic served by elegance and economy of modeling; vividness and power of imagination; and accuracy and realism of perspective.

What, though, is the *realism* of strategic analysis? The strategic analyst is not seeking finally to be realistic in a metaphorical way. The truth which the strategic analyst pursues is not directly that of the novelist or the painter or the moviemaker. The strategic analyst is not explicitly attempting to capture the spirit and character of his times. Knowing that he cannot "fully" succeed, he is attempting to render a literally accurate picture of the future through basically an imaginative ability to characterize the dynamics of societies, countries, forces and decision makers. But certainly the aesthetic of strategic analysis will be *similar* to the aesthetic of the novelist or the playwrite or the moviemaker: the strategic analyst must have a sense of process in human events; a sense of the impact of personality and decision on outcomes; a sense of the limits that time, space, psychological and cognitive constraints place on possibilities and developments.

The crucial need for measures of real effectiveness becomes evident when we ponder the aesthetic of strategic analysis. Strategic analysis as a genre, a distinct enterprise, a tradition with its practitioners working in an evolving community, can develop only if there are standards of success, measures of achievement. In the future of strategic analysis there will evolve different analytic routines and techniques, different approaches and methods for achieving effectiveness. It must be possible in some sense to judge the validity of these. It must be possible to distinguish those techniques and methods which might be indispensable for long periods against most problems; and those which happen to apply at given historical moments for given problems.

Of course the most important motive for becoming a strategic analyst is the adventure. It is an inevitable, timeless endeavor, for all generations will face unprecedented, novel situations which must be understood and dealt with. By definition each problem of strategic analysis that emerges in the flow of time will be new. Strategic analysis does *not* resemble various art forms and sports in which the bounds remain relatively invariant, and the evolutions in technique and concept occur within a constrained framework. Strategic analysis will be the pursuit of a quarry — the future, really — in which each new foray, expedition or exploration will be a new challenge.

Certain problems in strategic analysis may arise from difficulties in applying new techniques to new problems. For example, cases of success in strategic analysis may not constitute bases for analyzing new problems: we will not want to analyze tomorrow's problems with yesterday's techniques where inappropriate. We might therefore anticipate a *fallacy of universality*. This fallacy involves the assumption that techniques and methods successful at a given time for a given problem under certain circumstances are applicable to a variety of other problems.

Another potential fallacy is that of *analysis for the sake of analysis*. Analysts will develop very vivid, compelling and elegant models in which

THE STRATEGIC ANALYST IN THE FUTURE: SPECULATIONS

they have invested much; yet many of these models will turn out to be unrealistic. Predictably some analysts will become more interested in the modeling process, the analytic endeavor itself, than in realism. Obviously, however, the process cannot become an end in itself. Nor can the process be justified on the basis of anything other than the hard quest for literal realism, even though the quest can never be totally successful.

As suggested above, another difficulty in strategic analysis is the ready skepticism of many that we will ever project future events with any meaningful degree of specificity and realism. Many will feel that if we construct models of the future which seem vivid, concrete, compelling and to have verisimilitude, we will be deceiving ourselves because we will not be able to understand through our imaginations enough about the actual processes in the future to arrive at anything close to a realistic point of view. This can only be overcome by communities of analysts, supported with cognitive technology, and the accumulation of expertise and skill, particularly through the extrasomatic *post mortem* memories which allow us to understand how to improve and hence to learn. There is really no acceptable alternative to the quest, but belief is required.

> ... There is no more delicate matter to take in hand, nor more dangerous to conduct, nor more doubtful in its success, than to set up as a leader in the introduction of changes. For he who innovates will have for his enemies all those who are well off under the existing order of things, and only lukewarm supporters in those who might be better off under the new. This ... results ... from the incredulity of mankind, who will never admit the merit of anything new, until they have seen it proved by the event.
>
> —Machiavelli

Another formidable obstacle to strategic analysis, at least in the near term, will be the acceptance of cognitive technology. Indeed, for anyone who accepts responsibility for facilitating technical innovation, the problem of developing strategies of successful transfer of technology from the laboratory to the operational world is almost always formidable. This clearly is the case with cognitive technology. The innovation manager faces a specific dilemma: an innovation should be proved constructive before it is integrated into an operational setting, yet the most credible assessment is provided by usage in an operational setting. I need not belabor the obvious by providing a dissertation on the already well-known sources of inertia which often constrain the transfer of technology and sustain the dilemma.

Yet I remain optimistic about the prospects for transferring cognitive technology. Perhaps I can suggest one important reason for my optimism by relating an experience. A friend who happened for awhile to be in charge of all analysts at a large government analysis center recently advertised in the local papers for two analyst positions. Over 400 applications were received. Two persons in their twenties, with PhD's, hard and soft science training, and familiarity with computer operations, were selected after much deliberation over many promising candidates. Such analysts would find working *without* machines to be unthinkable.

Strategic analysis is clearly a field for unusually talented, specially trained people whose chief incentive is the pursuit of understanding. Fortunately, there are ample numbers of these people. They are the major hope of strategic analysis. For as I have said, what will help us most in the long run are not consultants and expert observers but innovative practitioners among the analysts themselves. Certainly a large part of that innovation will be directed toward refinements of cognitive technology.

Here is an excerpt from a further imaginary conversation between two young strategic analysts a few years from now. Lorraine Miyakawa and Karl Edwards are having lunch in a cafeteria in one of the major government departments.

"Last Sunday," says Lorraine, "I went with Stephanie and Jennifer to the exhibition of surrealist paintings at the National Art Gallery. Do you know anything about the Surrealists? Their philosophy and theories?"

"Not really," says Karl.

"They believed in the surreal, that which is higher or above reality. Dreams. Output from the unconscious. Freud was big then. And they didn't like the daily world. They were already sensitive to the media in the early years of the century. We're hardened to it. We've adapted. But they suffered culture shock, used terms like 'cacophony.' So they retreated into dreams. They had places, apartments, set aside for dreaming. And the imagery in their paintings is dreamlike. It violates logic, physics,

THE STRATEGIC ANALYST IN THE FUTURE: SPECULATIONS 159

what seem to be the actual relationships of things. There are watches that bend like cheese and people whose bodies are part furniture. Some of it seems beautiful to me. But one of the things they believed was that by putting together strange combinations of things the element of chance would create higher meaning, higher truth. Am I boring you?"

"Not yet. Keep trying."

"Well it occurred to me that in a way the Surrealists are of interest to us. They were certainly concerned with the possible and the impossible. They were concerned with a kind of method of analysis or creation — and you and I create — which was deeply intuitive. I guess it was founded on faith in the unconscious. Yet if they weren't believers in reason and science, their paintings were often terribly realistic in their details. The watch that bends like cheese looks exactly like a watch in all other respects. You're wondering where I'm going with this, aren't you?"

"The suspense is building, I must admit," he says.

"I'm going somewhere, really. Now: we must have the machine memory, right? Slo-mo summaries of our changing views. Telescope our chains of analysis and all that. But as you say, it's all so incomplete now. I can really imagine how great it must have looked in the early Eighties. But now it's obvious that DFNs, futures and all that stuff doesn't capture it."

"Pure Stone Age."

"Right. We've discovered higher hills. But anyway, what you were talking about earlier, that you still haven't got a good enough picture of your context after you put together all the formats, is really relevant here. In other words, what we need is somehow to be able to produce better visual representations — better symbology — of our imaginative scenarios. Part of the time we try to be extremely rational and quote realistic unquote. We're using quote disciplined qualitative methods unquote. Actually, most of the time we just let our imaginations romp a bit."

"Sometimes for days at a time."

Lorraine laughs. "But even when we *think* we're being super rational, highly rigorous, verging on the scientific, we know we're going to be idiots much of the time. Right?"

"No argument."

"So what I'm saying is that we need some better way to draw those watches of ours we think are so damned realistic and then see when and where they bend a little. In other words, all our scenarios are going to be in various stages of surrealism. We're going to depict *some* processes with ridiculously unreal time frames. We're going to imagine decision making sequences that horrendously violate cultural, psychological or other constraints. We're going to establish relationships of cause and effect, many of which are pretty realistic mixed in with some that are ridiculous. If we do 'paintings' of a sort, maybe they'll be realistic, impressionistic, and surrealistic in various parts all at the same time. And yet, just by having let ourselves go, we might also become more accurate in our estimates."

"Yes," says Karl, "but there's a trap here. I see where you're going, but be careful."

"What's the trap?"

"Just as beauty is in the eye of the beholder, in this case truth or realism or accuracy is also in the eye of the beholder. So how do we recognize problems?"

"You mean how do we see we're surrealistic when we already think we're realistic?"

"Exactly."

"It may not be hopeless. For one thing others must be able to review what we do and recognize the implausible even when we're blind to it. Somehow PFMs, DENs, and so on don't lend themselves to such overviews. They don't help the recognition problem. But put that aside for awhile. Bear with me and let's get back to that faith the Surrealists had in themselves. And let's avoid, to begin with, any judgments about how much more realistic and important we are than they were. Or that they were corrupt or irresponsible and we're not. We have our secret wishes in what we're doing just as they did."

"I'm humble."

"OK. So this faith the Surrealists had in themselves: Though there are obvious differences, we need something like that faith too. And here's a possible foundation for ours. First of all, we accumulate enormous experience bases in our lives. Actually we witness many of the basic processes of the world hundreds of times in one guise or another during our lives. Maybe in some ways there is less new under the sun than we imagine, although the *guise* changes often. Besides, in one sense it takes eons for things to evolve, to change; a century, an era, is only a microsecond. So in a sense, we shouldn't overly stress future shock."

"Believe me, I wouldn't."

"Now, here's the crux: If we can find ways to dramatize powerfully our depictions of future processes, so that their rationales, their assumptions, are at least implicit in the very structure of the depiction (and, ideally, explicit, flamingly obvious) and record this in the machine as we go about doing the analysis, we *do* have a chance, *if* you keep faith that we really know more than we think about the processes of the world. If you believe we've learned a lot and have it stored, then the problem becomes the recognition problem. Then there are two questions: One, can or will we recognize surrealism? Two, if so, when? May I keep going?"

"Please."

"First, if *we're* blinded at a given time, somebody else might see it on the superdepictions right away. Second, after a while — and this may be crucial — after a while, we may see it too. Maybe we'll just need some time for our data base to assert itself and for our initial excitement and possessiveness about our great insights to subside so the blinders come off and we sober up.

"And look: we haven't sacrificed that initial need to open up and freely romp for awhile, turn ourselves loose in analysis. Ultimately we don't pay a price. Secondly, maybe we're onto a sort of aesthetic unique to us — the recognition aesthetic. It's really a kind of aesthetic built around the learning problem. Which is certainly basic for us."

"Amen."

There is no use trying here and now to specify the ideal personality of the strategic analyst. But at least we must ponder that personality. Certainly the strategic analyst must live easily among opposites. He will embark on courageous adventures of the imagination, demanding free play and the conditions of inspiration; yet work under great discipline against ultimately insuperable odds, seeking the most difficult of perfections. Often he will need to exercise great patience in an atmosphere of urgency and deadlines. He will work hard within rather standard rules and frequently in conventional, impersonal settings populated by machines (though not always if such ingenious architects of the man-machine "interface" as Professor Nicholas Negroponte of MIT have their way). In whatever setting, however, the analyst must imagine unprecedented situations and events. Somehow he must avoid the unrealistic perspective and mentality of the bunker or the solipcism of the technocrat who lives in antiseptic worlds.

In short, the strategic analyst will be at war with many of his natural inclinations. There seems little doubt that he will have to learn to embrace new disciplines.

To borrow a description given to one of Miles Davis' new directions in music, strategic art ultimately becomes a bitch of a brew.

It seeks literal realism under the strategic imperative through the strategic imagination, taking the artist out of modern and postmodern inward adventures back not merely to the Present but to the next New World, the Future, a rebellious yet traditional thrust of great daring which, though there seems an inevitability about it, hardly promises safe passage and for a long time may inspire little, if any, confidence in other than a few. Meanwhile it makes the artist an apprentice to technology and a master of utilitarian machines, a disciplinarian informed by the rules of systems thinking but not bound by them, alliances which only a few years ago we might naively have thought impossible if not corrupt, even very dangerous, an abrogation of the necessary "primitiveness" or, depending on your view, "innocence" of the creative artist. Worse, it involves the artist directly in the decisions of politics, policy, economics and the rest.

Yet, really, he has always been there, or wanted to be there (if sometimes only under safe conditions), and should be there, as I have argued here. He will be there, too, increasingly, for the strategic adventure is unavoidable; it is a way back into the world, that home growing ever more dangerous and where survival in the forms we think we want is less and less likely to take care of itself.

As a final though on the *esprit* of strategic analysts, and with only a few reservations (after all, he was "mad"), there is sympathy with an exhortation made by Captain Ahab, one of the great challengers of the world's perversity:

> All visible objects, man, are but as pasteboard masks. But in each event — in the living act, the undoubted deed — there, some unknown but still reasoning thing puts forth the mouldings of its features from behind the unreasoning mask. If man will strike, strike through the mask!

Source Notes

1. INTRODUCTION

Page

(1) Yet obviously strategic analysis: A number of impressive and important writings on the issue of military warning in the international arena are useful to consult with respect to some of the difficulties in strategic analysis. These works include: Roberta Wohlstetter, *Pearl Harbor: Warning and Decision*, Stanford, California: Stanford University Press, 1962. Barton Whaley, *Codeword Barbarossa*, Cambridge, Massachusetts: MIT Press, 1973. Avi Schlaim, "Failures in National Intelligence Estimates: The Case of the Yom Kippur War," *World Politics* 28 (April), pp. 348-380, Thomas G. Belden, "Indications, Warning, and Crisis Operations," *International Studies Quarterly*, Vol. 21, No. 1, March 1977, pp. 181-198.

(3) Richards J. Heuer, Jr.: Richards J. Heuer, Jr., "Cognitive Biases: Problems in Hindsight Analysis," *Intelligence Studies* XXII/2, Summer 1978, pp. 21-28.

(4) and other scientists such as Amos Tversky, Richard Nisbett and Lee Ross: I will describe portions of their work in later chapters. Specific writings will be cited in the discussion.

(5) certain approaches developed by Willis Harman: Methodologies of forecasting developed by Harman are described in Chapter 4.

2. STRATEGIC ANALYSIS MODEL

(14) We may distinguish four types of projections: these distinctions are not new, though the terminology and rationale are. The difference between unimpeded and influenced projections bears on theoretical aspects of the nature of crisis, for example, the concept that crises and their management involve significant — perhaps decisive — *interactions* between and among parties. (See Belden, *op. cit.* and Charles McClelland, "Anticipation of Crises," *International Studies Quarterly*, Vol. 21, No. 1, March 1977, pp. 15-38, for representative discussions.

(16) For each of the four types of projection: I am indebted to Thomas Belden for insightful discussions pertaining to the uses of the information categories in developing formats for analytical reportage and, more basically, for strategic analysis itself. See also Belden, *Ibid.*

3. STRATEGIC ANALYSIS MEASURES

(24) Portions of James Grier Miller's work: see James Grier Miller, *Living Systems*. New York: McGraw-Hill Book Company, 1978. I am indebted to James Miller for discussing with me portions of his work and their utility to the present research.

(26) "Meaning represents": Miller, *Living Systems*, p. 11.

(29) Skeptics of *post mortems*: See Heuer, *op. cit.*

(30) These include the assumption that usually multidimensionality is good in strategic analysis: this wise assumption seems widespread.

(30) Another major assumption (concerns) safeguard procedures: this too is widespread. A fine discussion of safeguards may be found in Avi Shlaim, *op. cit.*

SOURCE NOTES 165

(31) "We tell ourselves": See Joan Didion, *The White Album*. New York: Simon and Schuster, 1979, p. 11.

(32) "Even though human behavior": See Edward O. Wilson, *On Human Nature*, Cambridge, Massachusetts: Harvard University Press, 1978, p. 73.

(32) classification of the specific obstacles to analytic effectiveness. I am indebted to Shlaim, *op. cit.*, for his fine discussion of some of the pitfalls in warning analysis. Relevant perspective also exists in a book by the Soviet authors, V. Druzhinin and D. Kontorov, *Concept, Algorithm, and Decision*, translated and published under the auspices of the United States Air Force, U. S. Government Printing Office, Stock Number 0870-00340.

(33) In Soviet Literature: see Druzhinin and Kontorov, *Ibid.*

(33) As Shlaim in particular discusses: see Shlaim, *op. cit.*

(35) Belden, *op. cit.*, p. 191.

(35) learning from "ambiguous" and "incomplete" information: a concept discussed broadly by Druzhinin and Kontorov, *op. cit.*

(35) Procedures/techniques for weighing opposing hypotheses: several writers discuss briefly this fundamental strategy of warning analysis. See Wohlstetter and Shlaim.

(35) Procedures that foster systematic use of a mixture of analytic approaches: Willis W. Harman has written valuably on this analytic strategy in the context of forecasting. See Willis W. Harman, *An Incomplete Guide to the Future*, New York: W. W. Norton and Company, 1979, pp. 9-19.

(39) To derive real and estimated effectiveness: the overall discussion of information-related variables draws in a very general sense from some of Miller's conceptualizing in *Living Systems*, especially those sections dealing with information subsystems in the individual and the group. See particularly Chapters 8 and 9.

(40)   The *relationality* variable: insofar as I know, this variable is unique to the present conceptualization of the strategic analysis process. That is, it is a new concept with respect to the measurement of strategic analysis. Of course, the notion of organized procedure is hardly new.

(41)   The most important of the three variables, signification: insofar as I know, this variable also represents a new conceptualization with regard to strategic analysis. The persistent problem of meaning, of course, is essentially bound-up with the signification variable.

(41)   The Soviets distinguish: see Druzhinin and Kontorov, *op. cit.*

(42)   Richards J. Heuer, Jr., has recently reviewed: The discussion of hindsight problems draws on Heuer, *op. cit.* Sources Heuer draws on include: Fischoff, Baruch and Beyth, Ruth. "I Knew It Would Happen: Remembered Probabilities of Once-Future Things," *Organizational Behavior and Human Performance*, 13 (1975); Fischoff, Baruch, *The Perceived Informativeness of Factual Information*, Technical Report DDI-1, Oregon Research Institute, Eugene, Oregon, 1976; Fischoff, Baruch, "Hindsight ≠ Foresight: The Effect of Outcome Knowledge on Judgment Under Uncertainty," Journal of Experimental Psychology: Human Perception and Performance, I, 3 (1975).

(43)   Carl Sagan (has) stressed: see Carl Sagan, *The Dragons of Eden*. New York: Random House, Inc. 1977. See especially Chapter 8:

(44)   appointed committees who explore an analysis failure: A review of some *post mortems* conducted in relation to the U.S. Intelligence Community is to be found in Richard W. Shryock, "The Intelligence Community Post-Mortem Program, 1973-1975," *Intelligence Studies* XXI/2, Summer, pp. 15-28.

(46)   "a configuration within the brain": see Wilson, *op. cit.*, pp. 75-76.

(47)   We can now estimate: see Sagan, *op. cit.*, especially Chapter 2.

(55)   as Richards J. Heuer, Jr., has remarked: the comments are contained in a memo to the present author.

SOURCE NOTES 167

(59)   In a recent book of considerable importance: the discussions on pp. 59-64 draw very heavily from the superb book on human inference by Richard Nisbett and Lee Ross. See Richard Nisbett and Lee Ross, *Human Inference: Strategies and Shortcomings of Social Judgment*. Englewood Cliffs, New Jersey: Prentice-Hall, Inc., 1980. This book seems to me one of the most important statements to come out of recent research in cognitive psychology. I base important portions of the conceptualizing in Chapter 3 on the work of Nisbett/Ross. The work of Paul Slovic, Baruch Fischoff and Sarah Lichtenstein, together with that of Richards Heuer, also have influenced me considerably. Heuer's work is cited in various places in the text. With respect to Slovic, Fischoff and Lichtenstein, see especially "Behavioral Decision Theory," *American Review of Psychology*, 28:1-39, 1977.

(61)   "the policy": Nisbett/Ross, *op. cit.*, p. 42.

(61)   "Causal analysis": Nisbett/Ross, *Ibid.*, p. 137.

(62)   "People do not seem": Nisbett/Ross, *Ibid.*, p. 165.

(62)   "Belief perseverence": Nisbett/Ross, *Ibid.*, p. 192.

(62)   "privileged epistemic": See an exchange between Adolf Grunbaum and Edward Rothstein, "How Valid is Psychoanalysis? An Exchange," *The New York Review of Books*, March 5, 1981, pp. 40-42.

(63)   "the persistence": Nisbett/Ross, *Ibid.*, pp. 247-8.

(65)   In the work of Hubert Dreyfus: see Hubert Dreyfus, *What Computers Can't Do*. New York: Harper, 1977, pp. 11-47. My discussion in the present book draws from some of Dreyfus' comments.

(67)   There are competing theories: For extended discussion, see especially: E.D. Hirsch, *The Aims of Interpretation* and *Validity in Interpretation*; Frank Kermode, *The Genesis of Secrecy: On The Interpretation of Narrative*; any number of works concerning New Critics and New Criticism, a powerful movement in the literary world influential especially in the 1940's; Hirsch's review of Kermode's *Genesis*, "Carnal Know

ledge," *The New York Review of Books*, June 14, 1979, pp. 18-20; and Gerald Graff's review of *Genesis*, "The Genesis of Secrecy: On the Interpretation of Narrative," *The New Republic*, June 9, 1979, pp. 27-32.

(77) pathological states in information systems: see Miller, *op. cit.* pp. 81-82 for a general discussion; see also pp. 473-480 for a discussion bearing on the individual human; and pp. 581-592 for consideration of these states in groups. See also pp. 121-202 for an extended consideration of the problem of information input overload, including a survey of related research, a study of the pathological effects of information input overload, and the results of experiments conducted by Miller and associates.

(78) there is a set of hypotheses about the likely behavior of information-related systems under varying conditions: see Miller, *Ibid.*, for an extremely detailed treatment of such hypotheses, related to his structure of information-processing subsystems in living systems. The individual discussions appear at many points in *Living Systems*.

(79) "What happens appears to be": Didion, *op. cit.*, p. 44.

(80) "In this light all narrative": Didion, *Ibid.*, p. 13.

4. THE ART OF STRATEGIC ANALYSIS

(86) "The Early Warning and Monitoring System": see Judith Ayres Daly, *Command, Control, and Intelligence: R&D for Decision and Forecasting Systems*. Arlington, Virginia: Cybernetics Technology Office, Defense Advanced Research Projects Agency, 1979, p. 12. For related and expanded descriptions, see the following: S. J. Andriole, *Progress Report on the Development of an Integrated Crisis Warning System*, Decisions and Designs, Inc., McLean, Virginia, December 1976; J. A. Daly and T. R. Davies, *The Early Warning and Monitoring System: A Progress Report*, Decisions and Designs, Inc., McLean, Virginia, July 1978.

(87) "Man is the model-making organism *par excellence*": see Edward T. Hall, *Beyond Culture*. Garden City, New York: Anchor Press/Doubleday, 1976, p. 10.

(87) *Situations* are considered as "things": V. Druzhinin and D. Kontorov, *op. cit.*, p. 103, agree that situations are of primary importance to intelligence analysts. See also Hall, *op. cit.*, pp. 113-123.

(88) We now turn to specific examples of analytic forms: Don R. Harris and Frances M. Calaway, two colleagues, have contributed significantly to the design of some of the analytic forms described and discussed in the present study. Harris has been involved primarily with forms used in the threat recognition stage; Calaway with forms used in the projection stage.

(98) Consider that the strategic analyst undertakes to develop scenarios: see Albert Clarkson, "Writing and Editing Scenarios," *Journal of Technical Communications*, Volume 16, No. 2, Second Quarter, 1969, pp. 20-21. Clarkson's article is drawn upon for the present discussion on scenarios.

(100) The anthropologist, Edward Hall, has written importantly about the relation of context to meaning: see Hall, *op. cit.*, especially Chapters 7, 8, and 9 (pp. 91-123).

(101) ". . . context . . . carries varying proportions of the meaning": Hall, *Ibid.*, pp. 75-76.

(105) There is no intent to provide a survey of various projection techniques: for an excellent examination of the literature, see Richard W. Parker, *Crisis Forecasting and Crisis: A Critical Examination of the Literature*. McLean, Virginia: Decisions and Designs, Inc., 1976.

(106) The format, first adopted by Belden: Belden, *op. cit.*

(111) In his very useful book . . . Willis Harman: see Willis W. Harman, *An Incomplete Guide to the Future*. New York: W. W. Norton and Company, 1979. I am indebted to Harman, and shamelessly

draw from his excellent discussion of the methods of futures research, especially pp. 9-19. In the present discussion, portions of the consideration of projection methodology are drawn exclusively from Harman and rely considerably on his perspective on methodology. All references to Harman in these pages refer to *An Incomplete Guide to the Future* and are part of the discussions therein on pp. 9-19.

(112) "behave like integrated": Harman, *Ibid.*, p. 14.

(112) "change is not aimless": Harman, *Ibid.*, p. 14.

(113) *Probable Futures Mapping*: A technique largely developed by a colleague, Frances M. Calaway.

(120) Thomas Belden has modeled this phenomenon as a decision stairway: Belden, *op. cit.*

(122) There are a number of approaches: see Parker, *op. cit.*

(123) "the elucidation, measurement and analysis": see Gerald W. Hopple, *Mapping the Terrain of Command Psychophysiology*. McLean, Virginia: International Public Policy Research Corporation, August 1978, p. 2.

(123) "As is customary in social scientific inquiry": Hopple, *Ibid.*, p. 3.

(124) Thomas Belden has described: Belden, *op. cit.*, p. 194.

5. THE STRATEGIC ANALYST IN THE FUTURE: SOME SPECULATIONS

(141) "If players are": See Frank Deford, *Big Bill Tilden*. New York: Simon and Schuster, 1975, p. 30.

(141) "He could rattle on": Deford, *Ibid.*, p. 102.

(141) "What Bird did": See Arnold Shaw, "Charlie Parker: Seque to Today," in *Esquire's World of Jazz*. New York: Thomas J. Crowell Company, 1975, p. 96.

(143) The physics of Newton: Mas'ud Zavarzadeh, in a superb study of new forms of American literature, has discussed the impact of recent and current physics, as well as modern communications media, on the novelist. See Mas'ud Zavarzadeh, *The Mythopoeic Reality*, Urbana, Illinois: University of Illinois Press, 1976, pp. 3-49. The general epistomological impact of both physics and the media has been considered by a number of observers, but Zavarzadeh addresses its specific effects on contemporary American literature illuminatingly. The discussion of these subjects in the present study has drawn for some of its points from the fine study by Zavarzadeh.

(144) "The behavior of the particle": see Cecil Schneer, *The Evolution of Physical Science*. New York, Grove Press, 1960, p. 364.

(144) "(George Smiley) hated the Press": John LeCarré, The LeCarré Omnibus: *Call for the Dead and A Murder of Quality*. London: Victor Gollancz Ltd., 1967, p. 126.

(145) "The new communication technologies": Zavarzadeh, *op. cit.*, p. 7.

(145) "the official level of reality": see *Cutting Edges: Young American Fiction for the 70's*, ed. Jack Hicks. New York: Holt, Rinehart, and Winston, 1973, p. 537.

(146) reality . . . is "neither significant nor absurd.": see Alain Robbe-Grillet, *For a New Novel*. New York: Grove Press, 1965, p. 19. Cited by Zavarzadeh.

(146) In a recent study: Zavarzadeh, *op. cit.*

(146) "Mistrust and even fear": Zavarzadeh, *op. cit.*, pp. 7-8.

(147) " . . . a form of organized chaos": Zavarzadeh, *Ibid.*, p. 9.

(157) "There is no more delicate matter": see Niccolo Machiavelli, *The Prince*. New York: The New American Library of World Literature, Inc., 1952, p. 55.

(162) "All visible objects": from *Moby Dick*.

Selected Bibliography

Allison, Graham T., *Essence of Decision: Explaining the Cuban Missile Crisis*. Boston: Little, Brown, 1971.

Andriole, Stephen J., *Progress Report on the Development of an Integrated Crisis Warning System*. McLean, Virginia: Decisions and Designs, Inc. Sponsored by Cybernetics Technology Office, Defense Advanced Research Projects Agency, December 1976.

Belden, Thomas G., "Indications, Warning, and Crisis Operations." *International Studies Quarterly*, Volume 21, Number 1, March 1977, pp. 181-198.

Ben-Zvi, Abraham, "Hindsight and Foresight: A Conceptual Framework for the Analysis of Surprise Attacks." *World Politics*, 28 April 1976, pp. 381-95.

Chan, Steve, "The Intelligence of Stupidity: Understanding Failures in Strategic Warning," *The American Political Science Review*, Volume 73 (1979), pp. 171-180.

Clarkson, Albert, *On Developing an Analyst-Oriented I&W System*. Sunnyvale, California: ESL Incorporated, N736, Spring 1977.

Clarkson, Albert, "Writing and Editing Scenarios." *Journal of Technical Communications*, Volume 16, Number 2, Second Quarter (1969), pp. 21-24.

Clauser, Jerome K. and Weir, Sandra M., *Intelligence Research Methodology*. State College, Pennsylvania: HRB Singer, Inc., 1976.

Daly, Judith Ayres, *Early Warning and Monitoring Prototype System: Sample Output.* McLean, Virginia: Decisions and Designs, Incorporated, January 1978.

Daly, Judith Ayres and Davies, Thomas R., *The Early Warning and Monitoring System: A Progress Report.* McLean, Virginia: Decisions and Designs, Inc., July 1978.

Daly, Judith Ayres, *Command, Control, and Intelligence: R&D for Decision and Forecasting Systems,* Washington, DC: Cybernetics Technology Office, Defense Advanced Research Projects Agency, Technical Memorandum, 1979.

Didion, Joan, *The White Album.* New York: Simon and Schuster, 1979.

Franco, G. Robert, *Quantitative Methods for Long-Range Environmental Forecasting: Long-Range European Projections.* Arlington, Virginia: Consolidated Analysis Centers, Inc., Volume I, March 1974.

Galbraith, John Kenneth, "The Strategic Mind," *The New York Review of Books,* October 12, 1978, pp. 72-74.

Hall, Edward T., *Beyond Culture,* Garden City, New York: Anchor Press/Doubleday, 1976.

Hampshire, Stuart, "Human Nature," *The New York Review of Books,* December 6, 1979, pp. c-d.

Harman, Willis, *An Incomplete Guide to the Future,* New York: W. W. Norton and Company, 1979.

Harris, Don R., *A Proposed Approach to Systematic Intelligence Analysis,* Sunnyvale, California: ESL Incorporated, Technical Memorandum, April 10, 1978.

Heuer, Richards J., "Cognitive Biases: Problems in Hindsight Analysis," *Studies in Intelligence* XXII/2, Summer (1978), pp. 21-28.

Hofstadter, Douglas R., *Godel, Escher, Bach: An Eternal Golden Braid.* New York: Basic Books, 1979.

Hopple, Gerald W., *Mapping the Terrain of Command Psychophysiology.* McLean, Virginia: International Public Policy Research Corporation, Technical Assessment Report 78-2-1, August 1978.

Hopple, Gerald W., *Political Psychology and Biopolitics: Assessing and Predicting Elite Behavior in Foreign Policy Crises.* Boulder, Colorado: Westview Press, 1980.

Jervis, Robert, *The Logic of Images in International Relations,* Princeton, New Jersey: Princeton University Press, 1970.

Kahn, Herman and Weiner, Anthony J., *The Year 2000,* New York: The MacMillan Company, 1967.

Lederberg, Joshua, "Digital Communications and the Conduct of Science: The New Literacy," Preceedings of the IEEE, Vol. 66, Number 11, November 1978, pp. 1314-1319.

Mahoney, Robert B., *Executive Aid for Crisis Management: Sample Output.* Arlington, Virginia: Consolidated Analysis Centers, Incorporated, November 1977.

McIlroy, John J., *Developmental Methodologies for Medium- to Long-Range Estimates: Executive Summary.* Arlington, Virginia: Consolidated Analysis Centers, Inc., September 1976.

Miller, James Grier, *Living Systems,* New York: McGraw-Hill, 1979.

Omen, George E., *A Forecasting Model of International Conflict.* Unpublished PhD Dissertation, University of Hawaii, August 1975.

Robitschek, Fred W., "Comparative Definitions of Warning Elements," unpublished notes (1978-79).

Rothstein, Edward, "The Dream of Mind and Machine," *The New York Review of Books,* December 6, 1979, pp. 34-39.

Sagan, Carl, *The Dragons of Eden,* New York: Random House, Inc., 1977.

Schank, Roger C., *Research at Yale in Natural Language Processing.* New Haven: Yale University, Department of Computer Science, 1976.

Schlaim, Avi., "Failures in National Intelligence Estimates: The Case of the Yom Kippur War," *World Politics* 28 (April), pp. 343-380.

Shryock, Richard W., "The Intelligence Community Post-Mortem Program, 1973-1975," *Studies in Intelligence* XXIII/2, Summer (1979), pp. 15-28.

Sutherland, John W., "Towards a Middle-Range Theory of Societal Dynamics." Forthcoming in *Human Relations*.

Sutherland, John W., *Societal Systems: Methodology, Modeling and Management.* New York: Elsevier-North Holland, 1978.

Sutherland, John W., *Systems Analysis, Administration, and Architecture.* New York: Van Nostrand Reinhold Company, 1975.

Toynbee, Arnold J., *A Study of History* (Abridgement of Volumes I-VI by D. C. Somerville). London: Oxford University Press, 1946.

Wilson, Edward O., *On Human Nature.* Cambridge, Massachusetts: Harvard University Press, 1978.

Wohlstetter, Roberta, *Pearl Harbor: Warning and Decision.* Stanford, California: Stanford University Press, 1962.

Wohlstetter, Roberta, *Cuba and Pearl Harbor: Hindsight and Foresight.* Santa Monica: Rand Corporation. RM-4328, April 1965.

Zavarzadeh, Mas'ud, *The Mythopoeic Reality.* Chicago: University of Illinois Press, 1976.

Index

Actionability, 69
Ambiguity tolerance, 66
American Literature, 145
Analysis
    accuracy, 57
    art, 56, 84
    communities, 29, 43
    heuristics, 57
    routines, 96
    safeguards, 11
    strategies, 57
An Incomplete Guide to the Future
    (Harman), 111
Antibias techniques, 27
Artificial intelligence, 1, 52
Artificial schemata, 52, 88, 101
Associative functions, 25
Availability heuristic, 59

Backlog, 21, 39, 100, 103, 126, 129
Bayesian analytic approaches, 123
Belden, Thomas G., 35, 106, 120, 124
Belief perseverance, 62
Bias, 34, 97
Bohr, Niels, 144
Bureaucratic analytic arrangements, 11

Carlyle, Thomas, 107
Causal analysis, 61
Chunking data, 25
Cognitive fitness, 58
Cognitive obstacles, 33-34
Cognitive psychologists, 3
Cognitive technology, 1, 43, 83, 103, 134
Context, 101
Cultural data, 13

Daly, Judith Ayres, 86
Davis, Miles, 162
Deception, 34
Decider, 25
Decision maker, 16, 67
Decision stairway, 120, 128
Defense Advanced Research Projects Agency, 86, 123
Devil's advocacy, 54
Didion, Joan, 31, 79
Displays, 53
Dreyfus, Hubert, 52, 65

Early Warning and Monitoring System, 86
Economic data, 13

Economy of modeling, 69
Effectiveness
  definition of, 29
  estimated, 29
  obstacles to, 30-35
  problem of, 41-43
  real, 29
  viewpoint on, 68
Einstein, Albert, 143
Empiricism, 9
Epistemology, 2, 31-33
Event, 15
Extrasomatic memory, 2, 43, 46, 53, 64, 103

Forecast, 16
Forms of analysis, 83
Friendly decision maker, 18
Fringe consciousness, 65
Future-backward projection, 120
Futures mapping techniques, 113-15

Gestalt, 46, 65
Global information processing, 65
Grammar of projections, 16

Hall, Edward T., 87, 100-01
Harman, Willis, 5, 111
Heuer, Richards J., 3, 42, 55, 59, 61
Hiroshima, 33
Hopple, Gerald W., 123
Human Inference: Strategies and Shortcomings of Social Judgment (Nisbett/Ross), 59
Hydraulic view of causality, 62

Imagination, 98
Indicator, 10, 86-87
Information
  categories, 106-08
  decisions, 41
  machines, 1, 7
  overload, 3
  variables, 39
Interpretation, 31

Knowledge structures, 2, 46, 59, 61, 82
  See also Artificial schemata

Learning, 3
Le Carré, John, 144
Literature of inspiration, 62
Living systems theories, 24

Maintenance, 21, 26, 37, 71
Markovian analytic approaches, 123
Meaning, 24-25, 102
Memory, 42, 49
Methodological inflexibility, 34
Miller, James Grier, 5, 24
Mind set, 58
Mirror imaging, 97
Modeling, 31, 58
Monitoring stage, 11, 86, 135

Nagasaki, 33
Narratology, 31
Newtonian physics, 143
Nisbett, Richard, 4, 61
Novel situations, 14, 33, 94

Operational context, 3
Overreliance
  on personality, 34
  on theory, 34

Parker, Charlie, 141
Pathologies and information systems, 77
Pearl Harbor, 31

Post mortem, 4, 29, 36, 41-44, 55, 61, 65, 82, 124, 132
Prediction, 14, 16
Probability rules, 124
Projection stage, 14, 105
Psychophysiological constraints, 2

Quantification, 2
Queing data, 25

Rationale recording, 122
Readiness, 21, 26, 35-37, 74
Realism, 2, 9
Relationality, 21, 39-40, 75, 100, 103, 126, 129
Representativeness heuristic, 59
Restricted perspective, 33
Ross, Lee, 4, 61

Scenario, 87, 99
Schemata, 45-46, 52, 82, 96
Schlaim, Avi, 33
Sector comparison, 113
Signification, 21, 39, 41, 44, 75, 98, 100, 128, 130
Sisyphus, 31, 98
Situation, 87
Soviet military literature, 33
Specificity Rating Scale, 16, 108, 110, 126
Strategic Analysis
  definition of, 9-10
  measures of, 21-80
  model of, 7-19
  settings of, 10-11
  symbolic representation of, 47-54
Strategic analyst, 2, 7-8, 47-54, 141-50
Strategic art, 5, 81, 141

Strategic imagination, 4, 56, 64, 97
Surprise, 30
Sutherland, John W., 5, 69
Systems science, 5

Tilden, Bill, 141
Threat models, 15, 88-96
Threat recognition stage, 14, 87, 97, 137
Tolstoy, Leo, 107
Tversky, Amos, 4

Variables, 56
Verisimilitude, 10
Vividness factor, 60, 133

Warning, 10
Weather forecasting, 14
Whaley, Barton, 31
Wilson, Edward O., 46
Wishful thinking, 97
Wohlstetter, Roberta, 31

DATE DUE